Quantitative

适用于2023年改革后的GRE数学考试

解密GRE
数学出题点

程黛苑 ◎ 编著

全新改版

机械工业出版社
CHINA MACHINE PRESS

本书全面系统地梳理、归纳、讲解 GRE 数学考点。第一篇论述了 GRE 数学在 GRE 考试中的重要性，简单介绍了 GRE 数学的主要考查内容、GRE 数学考试题型和注意事项。第二篇详细剖析了数论、代数、几何、数据分析这四大考查内容，分析了每个考点涉及的概念和知识点在真正考试中的考查形式、考法和解题思路，并且配有若干例题和对应练习。每一道题都呈现第一视角的解题思路，并不只是列出公式、给出答案。第三篇提供了两套符合当前考试形式的数学模考练习题，帮助考生进一步熟悉相关题目的问法、解题的时间分配。附录部分提供了 GRE 数学相关的词汇和常用公式，帮助考生掌握、巩固相关数学术语和表达，以及熟练运用常用公式。

图书在版编目（CIP）数据

解密 GRE 数学出题点 / 程黛苑编著. -- 北京：机械工业出版社，2024.10. --（娓娓道来出国考试系列丛书）. -- ISBN 978-7-111-77032-9

Ⅰ. O13

中国国家版本馆 CIP 数据核字第 202478PL75 号

机械工业出版社（北京市百万庄大街 22 号　邮政编码 100037）
策划编辑：苏筛琴　　　责任编辑：苏筛琴
责任校对：张若男　　　责任印制：单爱军
保定市中画美凯印刷有限公司印刷
2024 年 11 月第 1 版第 1 次印刷
210mm×260mm · 19.75 印张 · 1 插页 · 545 千字
标准书号：ISBN 978-7-111-77032-9
定价：88.00 元

电话服务　　　　　　　　　网络服务
客服电话：010-88361066　　机　工　官　网：www.cmpbook.com
　　　　　010-88379833　　机　工　官　博：weibo.com/cmp1952
　　　　　010-68326294　　金　　书　　网：www.golden-book.com
封底无防伪标均为盗版　　　机工教育服务网：www.cmpedu.com

名师推荐
Recommendation

多年来，我本人有幸见证着程老师教学上的成长，她在教学、教研上的严谨和苛刻，她对自己的严格要求，都令我拜服。两次 GRE 语文、数学双满分的经历就是她能力与态度的双重证明。本书凝结了程黛苑老师十多年来的教学精华，它非常严格地覆盖了 GRE 数学部分官方的所有重要考点（数论、代数、几何、数据分析），选取了近些年真题当中具有代表性的例题，定能给备考 GRE 的同学提供全面、严谨的指导。

——万炜　GRE/LSAT/TOEFL 一线名师

GRE 数学冲击超高分对 GRE 总分取得理想目标来说至关重要！而现在的 GRE 数学出题灵活性是让我们不能小觑的。在本书中，程黛苑老师极其详细地梳理了 GRE 数学会考查到的所有考点，和实际考试中对应的各种考查方式。讲解浅显易懂、基础薄弱的同学也可以迅速学会并举一反三！相信这本书一定可以帮助到正在备考 GRE 的各位同学。

——闫晨晨　GRE/GMAT 阅读一线名师

很多中国同学觉得 GRE 数学可以轻松拿满分，所以在备考时没有给予足够的重视，导致考试失利。其实 GRE 数学中有很多难点和丢分点，需要大家系统准备并配合足够的练习去反思和总结。在本书中，Monica 老师将所有考点进行了系统梳理，针对不同题型配合了大量英文原题及讲解，大大提升同学们的备考效率。这本书不论是对数学基础薄弱的小白，还是对需要冲刺满分的高阶选手，都是备考 GRE 数学的不二选择。

——宋小明　GRE 填空一线名师

这本书涵盖 GRE 数学（定量分析）的所有知识点与考点，是一本与时俱进、翔尽全面的 GRE 备考书。不仅适于备战 GRE 考试，也能够帮助读者从中英文双语角度快速回顾基础数学知识，例如基本的统计学概念、排列组合问题、基本的数据分析方法等。充分解决了不知道中文语境下的数学概念如何用英文表达的问题。本书尤其适合计划攻读定量分析、计量、数据科学等与数学紧密结合的专业的同学作为基础参考书。

——潘晨光　GRE 填空一线名师，哥伦比亚大学在读博士

丛书序 Preface

"娓娓道来出国考试系列丛书"之"解密GRE系列"是青山学堂团队的经典力作，凝结了我们多年积累的教研成果，包含《解密GRE阅读逻辑线：双线阅读法》《解密GRE写作论证思维》《解密GRE填空核心考点：双线突破法》和《解密GRE数学出题点》。本丛书在上一版的基础上，针对2023年GRE改革进行了全面改版。

无论是我亲自动笔，还是邀请其他作者主笔，我一直坚持一个核心的理念，那就是拒绝"马后炮"式的方法论，要呈现第一视角的考场心态。在编写过程中，我们所有作者都把自己还原到正在考试的学生的角度，去感受考场上学生实际面对的思路上的困难，而不仅仅是呈现一个漂亮的答案解析。比如说，在GRE写作中，作为学生的我不想只看到老师给我呈现一篇思路完美的文章，因为老师想这个思路可能用了两个小时，最后写出的文章确实无懈可击，但在考场上我们根本没有这样奢侈的条件可以设计出这么精美的结构。再比如，在GRE阅读中，也许一道题选C，因为ABDE选项在文章中没有提过，但是C选项在原文的第七行出现了。真正的问题不是知道答案选项在第七行但看不出来，而是我当时不记得这个信息在哪里，所以我需要知道为什么第一遍阅读时应有第七行的意识，以及我看到C选项时应该能够回到第七行。毕竟，我读完文章后不可能记住所有的东西，我也不希望老师告诉我必须练到所有句子和词都记得住，我希望老师能够给我一套普适性强的方法论，教我在阅读时如何区分什么该理解，什么不需要关注。就像初中平面几何课程，我不希望老师告诉我在∠A上作辅助线可以解决某道题，我希望老师能够告诉我，当我看到这道题的时候，应该靠什么样的普适性强的方法论才能意识到应该在哪里作辅助线。

为了达到这个目的，我们要求作者们不仅仅讲出自己做题时第一视角的思路，还要更进一步深挖自己的直觉。因为，作为一个熟练的考生，很多习惯是下意识的——自己没有意识到自己其实已经做了某种思考，已经在潜意识里完成了这个步骤。我们都知道很多时候学霸给学渣讲题的时候会说"这很明显啊"，但是学渣感觉步骤并不明显。我们希望老师的讲解不要高高在上，不要把自己觉得想当然的步骤当作真正理所应当的。我们要求老师能够解构自己潜意识里的做题习惯，用清晰的语言呈现给读者。

本系列图书的另一个重要目的就是摒弃过去横行在出国考试培训当中的一系列"奇技淫巧"，一些从根本上来自于应试教育的习惯。这些习惯在高考和考研等考试当中也许有用，之前的一些培训从业者把它们用在了出国考试中，并且忽悠了无数学生，但其实这些习惯对于托福、GRE、GMAT 等考试毫无用处。比如在 GRE 阅读中，很多人宣称"原文没说的就不能选"；在 GRE 和托福写作当中，很多人倡导学生写长句，用"高级词"替换"低级词"……这些做法对学生不仅不会有帮助，实际上还会严重坑到很多学生。

总而言之，本系列图书试图以一种全新的视角把做题方法呈现给广大读者，力求做到新颖、诚实、细致、全面、可操作。祝愿考生们在拥有了这套利器之后能顺利攻克出国深造路上的层层难关。

万　炜
2024 年 10 月

前言
Foreword

本书针对的是正在备考 GRE、想要冲击 GRE 数学高分甚至满分的同学。

在 GRE 考试中，GRE 语文和数学的分值均为 170。在如今申请国外大学研究生越来越内卷的时代，很多学校过去要求总分 320 的标准现在可能已经变成了 325，甚至更高。如果考生在数学部分能取得一个诸如满分 170 的成绩，在很大程度上就能够缓解语文部分的备考压力。而且 GRE 数学部分相对于语文部分来说，提分效果通常会更迅速、更明显，所以数学高分往往是大家在准备 GRE 考试时能取得理想总分的必要条件。

目前市面上 GRE 备考资料鱼龙混杂，并不是所有资料都有价值，所以收集到的数学习题 ≠ 有效学习资料。GRE 数学比较合理的备考方式是：先通过了解整个 GRE 数学的知识体系和考点，知道每个知识点在真正的 GRE 考试中是如何考查的，以及我们要如何解题；然后再通过题目练习去巩固这些知识点和做题方法。要了解 GRE 数学考查的知识点，虽然可以参考《GRE 官方指南》(以下简称为 OG)，但是 OG 只是给予全球考生 GRE 考试达到全球平均水平的一份备考指南，而在全球数学平均分为 153、中国数学平均分为 166 的情况下，OG 上的很多知识点对于中国学生而言过于粗糙，往往缺乏实用性。只给了基本的定义，并没有详细讲解这些概念在真实考试中是如何被考查的，以及考生如何利用这些定义来解题；同时配套的练习题难度也远低于中国学生冲击目标分数所需要练习的难度。

为了解决上述问题，我结合了 10 年 GRE 的教学经验和 GRE 数学、语文双满分的做题思考，以及我所带出的高分学员之所以能取得高分甚至满分的各种因素，编写了这本书。本书除了针对重要概念给出基本定义，还会基于这些定义进一步拓展，让考生了解 GRE 考试中这些知识点的具体考查形式和解题套路，以便大家更好地举一反三。在练习题方面，从 2023 年 9 月 22 日 GRE 考试改革至今，我不断收集和整理改革后最新的数学机经真题，并从中选取了高频且中国考生易出错的题目，目的就是想给大家带来最优质、最贴近真实考试的 GRE 数学备考资料，以节约大家盲目找练习资料所消耗的时间。

这本《解密 GRE 数学出题点》有以下几个显著优势：

1. 全英文 GRE 考试原题

目前市面上很多流传的数学机经资料都是来自于考生的回忆，因此大部分题目都是中文题干或者中英文夹杂的形式，这样就弱化了 GRE 考生对英文读题能力的训练。而对于中国学生来说，读英文原题、理解题意是大家非常需要锻炼的能力之一。本书收录的全部是 GRE 数学全英文真题原题，能有效帮助大家锻炼 GRE 数学所需要的读题能力。并且为了让大家及时发现自己读题时候可能存在的误区，针对每道英文原题的题干，我都配上了相应的中文翻译。

2. 数学考点覆盖全面

我是长期从事 GRE 培训的一线名师，曾多次参加 GRE 考试，并多次取得 GRE 数学满分，甚至 GRE 数学、语文双满分的成绩，对于考试考查的真实情况非常了解。并且我在长达 10 年的 GRE 考试培训过程中，教授过上万名学生，对于广大考生的学习情况有非常科学的认知，对大部分中国考生的不足之处也了如指掌。本书为了贴合真正的考试情况，精细梳理了高频且考生易错的知识点，让考生发现薄弱点，有针对性地查漏补缺。同时选用源自于改革后 GRE 真实考试的原题，让大家有效了解 GRE 数学考试的新趋势，对最新考试难度也有更合理、更准确的认知。

3. 题目质量有保障

本书所有题目答案是由我整理和反复校对的，尽最大努力避免了答案和题干出错，消除了考生对题目正确性的困惑。一份优质的资料能让大家高效地备考，从而在复习规划上也更有针对性。

4. 配有详细题目解析

本书中的所有例题和练习题均配有详细解析。大家可以通过例题了解在考试中是如何考查各个章节知识点的，以及利用练习题来测试自己对上述知识点掌握得是否完善。

> **最后的叮嘱**
>
> 大家切忌高估自身的数学水平。很多论坛上至今还在盛传"中国学生的数学随便考考就是 169 分、170 分""GRE 数学很简单，考前刷刷机经就能拿满分了"这类不负责任的说法。然而实际情况是：从 2012 年 GRE 进入机考时代至今，GRE 数学题目的难度每年都在不断加大。
>
> 所以，大家在备考 GRE 的过程中，要正确认清自己的数学能力和水平，既不要妄自菲薄，也不要自视甚高，要根据自己 GRE 数学所需要的目标分数来合理安排 GRE 数学科目的备考投入时间，这才是最后拿高分的王道！

希望这本书能够帮助备考 GRE 的考生们提高备考效率，少走弯路。预祝大家取得满意的 GRE 成绩！

程黛苑
2024 年 10 月

目录 Contents

丛书序
前　言

第一篇　GRE 数学总论

一、GRE 数学在 GRE 考试中的重要性　　　　　　　　　　　　　　　/ 002
二、GRE 的考试形式以及数学的主要考查内容　　　　　　　　　　　/ 002
三、GRE 数学的题型　　　　　　　　　　　　　　　　　　　　　　/ 003
四、GRE 数学考试的注意事项　　　　　　　　　　　　　　　　　　/ 008
五、GRE 数学考试中的计算器　　　　　　　　　　　　　　　　　　/ 009

第二篇　GRE 数学考查的知识点

第一章　数论　　　　　　　　　　　　　　　　　　　　　　　　　/ 012

1.1　整数　　　　　　　　　　　　　　　　　　　　　　　　　　　/ 013
　　1.1.1　整数　　　　　　　　　　　　　　　　　　　　　　　　/ 013
　　1.1.2　因数和倍数　　　　　　　　　　　　　　　　　　　　　/ 015
　　1.1.3　最大公因数和最小公倍数　　　　　　　　　　　　　　　/ 019
　　1.1.4　商和余数　　　　　　　　　　　　　　　　　　　　　　/ 022
　　1.1.5　奇偶性　　　　　　　　　　　　　　　　　　　　　　　/ 029
　　1.1.6　质数和合数　　　　　　　　　　　　　　　　　　　　　/ 035
　　1.1.7　连续整数　　　　　　　　　　　　　　　　　　　　　　/ 043
　　1.1.8　数个数　　　　　　　　　　　　　　　　　　　　　　　/ 048
1.2　分数和小数　　　　　　　　　　　　　　　　　　　　　　　　/ 050
　　1.2.1　分数　　　　　　　　　　　　　　　　　　　　　　　　/ 050
　　1.2.2　小数　　　　　　　　　　　　　　　　　　　　　　　　/ 052

1.3 指数和根 / 064
　　1.3.1 指数 / 064
　　1.3.2 根 / 069
1.4 实数 / 072
1.5 比例/比率 / 074
1.6 百分比 / 078

第二章　代数 / 088
2.1 代数式运算 / 089
2.2 指数运算 / 094
2.3 根号运算 / 100
2.4 解线性方程 / 107
2.5 解一元二次方程 / 111
2.6 解不等式 / 115
2.7 数列 / 121
2.8 函数 / 128
2.9 函数图形 / 132

第三章　几何 / 134
3.1 直线和角 / 135
3.2 多边形 / 137
3.3 三角形 / 139
3.4 四边形 / 154
3.5 圆 / 165
3.6 立体几何 / 177
3.7 坐标几何 / 183

第四章　数据分析 / 193
4.1 描述数据的图表方法 / 194
4.2 描述数据的数值方法 / 200
4.3 统计方法 / 216
　　4.3.1 集合和数组 / 216
　　4.3.2 排列组合 / 225
4.4 概率 / 242

4.5	数据分布、随机变量和概率分布	/ 252
4.6	图表题	/ 255
4.7	应用题	/ 269
	4.7.1 平均数问题	/ 269
	4.7.2 物质混合问题	/ 270
	4.7.3 路程问题	/ 270
	4.7.4 生产问题	/ 274
	4.7.5 投资问题	/ 275
	4.7.6 钱的计算问题	/ 279

第三篇　模考套题训练

| 模考一 | / 282 |
| 模考二 | / 290 |

附　录

| 附录一　GRE 数学英文表达及词汇 | / 300 |
| 附录二　GRE 数学常用公式 | / 304 |

第一篇

GRE 数学总论

一、GRE 数学在 GRE 考试中的重要性

GRE 成绩由 Verbal Reasoning（语文）、Quantitative Reasoning（数学）和 Analytical Writing（写作）三个部分构成，其中语文和数学的满分均为 170 分，写作的满分为 6 分。在申请学校的时候，中介要求的 GRE 总分（如 320, 325, 330）特指的就是语文和数学的总分（不含写作成绩）。根据 ETS 2022 年公布的中国考生的数据，2022 年中国考生的 GRE 语文平均分为 153.4（全球 GRE 语文平均分为 151.3），2022 年中国考生的 GRE 数学平均分为 165.9（全球 GRE 数学平均分为 157）。中国学生从小就拥有良好的数学功底，在实际备考过程中，数学部分相对语文部分来说，也更容易提分，所以在备考的时候，拿到一个数学高分，能够在很大程度上缓解考生在语文部分的压力，因此取得一个数学高分甚至满分的成绩是考生 GRE 取得理想总分的重要保障。

 二、GRE 的考试形式以及数学的主要考查内容

Section	题量	时间
写作	一篇 Issue	30 min
语文 Section 01	12 道题	18 min
数学 Section 01	12 道题	21 min
语文 Section 02	15 道题	23 min
数学 Section 02	15 道题	26 min

或者

Section	题量	时间
写作	一篇 Issue	30 min
数学 Section 01	12 道题	21 min
语文 Section 01	12 道题	18 min
数学 Section 02	15 道题	26 min
语文 Section 02	15 道题	23 min

写作部分过后，考试顺序可能是语文、数学、语文、数学，也可能是数学、语文、数学、语文。

在考试中，GRE 数学部分共 27 道题，第一个部分 12 道题，21 分钟；第二个部分 15 道题，26 分钟，总题目数量为 27 道，总时长为 47 分钟。GRE 数学两个部分采用了自适应模式，即第一个部分的作答表现会决定第二个部分题目的难度。在各部分做题时间没有用完的情况下，考生还可以自由跳转到该部分的任意一道题目去更改答案；考生可以用 Mark 键标记自己不确定或者不会做的题目，然后点击 Review 键来查阅自己标记过的题目。并且在界面上点击 Calculator 键，会跳出系统自带的计算器，考生可以用它做一些数值计算。

在考试中，GRE 数学主要考查考生：

- 基本的计算能力
- 对基础数学概念的理解能力
- 建立数学模型以及用量化的方法解决问题的能力

考试中某些问题是在现实生活场景下提出的，某些是在纯数学场景下提出的。主要考查以下四大板块：

- Arithmetic 数论
- Algebra 代数
- Geometry 几何
- Data Analysis 数据分析

数论部分主要考查整数的性质，如整除、因数分解、质数、余数和奇偶性等，主要是国内小学阶段学的知识点。代数部分主要考查具体的运算，如解方程、解不等式、指数运算等，是初高中阶段学的知识点。几何部分主要考查几何知识、如平面几何、立体几何、解析几何，也是初高中阶段学的知识。数据分析考查的知识点比较多，主要是高中阶段的集合问题、排列组合和统计问题等。

由此可知，GRE 数学考查的是高中及高中之前的数学知识点，并不考查高等数学，如微积分，并且也不考查各种数学性质的证明。

GRE 数学默认有两个前提：①所有的数字都是实数；②几何图形不一定是按比例绘制的。

中国考生普遍在数论部分和数据分析部分失分较多。因为数论虽然是小学所学的知识点，但更偏向于小学奥数，考法非常灵活；而数据分析涉及的排列组合、统计部分对于数学基础不够扎实的同学来说可能有一定难度，并且读图表的能力也是中国学生普遍欠缺的。

三、GRE 数学的题型

Quantitative comparison questions 数量比较题
Multiple-choice questions—Select one answer choice 单项选择题
Multiple-choice questions—Select one or more answer choices 不定项选择题
Numeric entry questions 数字填空题

每个问题要么是一个独立的问题，要么是数据分析的一组问题中的一部分（这类问题我们通常称为图表题，考试的时候这类题会基于表格或图表提出 3 个问题）。在考试中，考生可以用屏幕上提供的计算器辅助计算（计算器的详细信息会在之后介绍）。

我们先来看一下每类题型的特点以及解题技巧。

1. 数量比较题

O is the center of the circle, and the perimeter of △ROS is 6.

Quantity A	Quantity B
The circumference of the circle	12

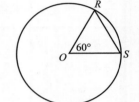

○ Quantity A is greater.
○ Quantity B is greater.
○ The two quantities are equal.
○ The relationship cannot be determined from the information given.

如上题所示，这类题目会让你比较两个数量——Quantity A 和 Quantity B，然后决定哪个选项描述了正确的大小关系。这类题会有 4 个固定的选项表达：

Quantity A is greater.（数量 A 更大。）

Quantity B is greater.（数量 B 更大。）

The two quantities are equal.（数量 A 与数量 B 相等。）

The relationship cannot be determined from the information given.（无法从题目信息判断数量 A 与数量 B 的大小关系。）

题干翻译

O 为圆的圆心，△ROS 的周长为 6。

解题思路

题目要求比较圆的周长和 12 的大小关系。

根据题干可知，O 为圆心，并且由图可知 $\angle ROS = 60°$，可以判断出 △ROS 为等边三角形。

△ROS 周长为 6，因此 $RO = OS = RS = 2$。

即圆的半径长度 = 2，

则圆的周长 = $2\pi r = 4\pi > 12$，

因此，数量 A 更大。

▶ **数量比较题的解题建议：**

（1）熟悉答案选项。如前文所述，数量比较题的选项是 4 个固定的表达，所以要熟记，以节约读题时间。尤其是最后一个选项"无法从题目信息判断数量 A 与数量 B 的大小关系"，如果数量 A 与数量 B 的数值是可以通过计算来确定的，那么就千万不要选择最后一个选项。此外，当我们确定一个数量大于另一个数量，请确保选择相应的选项，避免选择颠倒大小关系的选项。

（2）避免不必要的计算。不要为了比较两个数量而浪费时间进行不必要的计算。我们只需要做对确定大小关系有帮助的计算，过程完全可以简化，也可以利用转换或估值。

（3）注意，几何图形不一定是按比例绘制的。如果题目对于某一特定的几何图形的某些方面没有给出确定信息，我们应该尝试重新绘制该图形，保持题目给出的确定信息不变，通过改变图形中未确定的信息来检查大小关系是否发生了变化。

（4）代入数值。如果比较的数量中存在代数表达式，我们可以利用简单的数值代替变量，来分析比较结果。在给出答案之前，尽量考虑各种合适的数值，例如，零、正数和负数、比较小的数值和比较大的数值等。注意，代入数值只能帮我们确定 D 选项，无法保证其他选项的正确性。

2. 单项选择题

例题

If x and y are positive integers and $x + y = 8x + 22$, which of the following must be true?

○ x is even.   ○ xy is odd.

○ $x - y$ is odd.   ○ $x(y+1)$ is even.

○ x and y are both odd.

如上题所示，这类题目会有 5 个选项，要求考生从 5 个选项中只选出一个选项。

识别方式：在实际考试中，单项选择题前面的选项是椭圆形的。

题干翻译

如果 x 和 y 为正整数，$x + y = 8x + 22$，以下哪个说法一定是正确的？

解题思路

由题干可知，$x + y = 8x$（偶数）$+ 22$（偶数）$=$ 偶数。

因此，x 与 y 同奇或者同偶。

选项分析

由于 x 可能是奇数，也可能是偶数，所以第一个选项不符合要求；

xy 可能是奇数，也可能是偶数，所以第二个选项不符合要求；

$x - y$ 是偶数（同奇或同偶相加减，结果为偶数），所以第三个选项不符合要求；

如果 x, y 都是奇数，则 $y+1$ 为偶数，那么 $x(y+1)$ 为偶数；如果 x, y 都是偶数，那么 $x(y+1)$ 为偶数，因此第四个选项正确。

由于 x 与 y 同奇或者同偶，因此第五个选项不符合要求。

综上，答案为第四个选项。

▶ **单项选择题的做题建议：**

注意：答案一定在给出的选项中。如果你的答案不是给出的五个选项之一，那么你的答案就一定是出错了，建议执行以下操作：

（1）仔细重读问题——检查是否遗漏或误解了某些重要的信息。

（2）检查你的计算——看看是否存在计算失误，比如在使用计算器时出现了数字录入的错误。

（3）重新评估解题方法是否正确。

（4）对于需要取近似值的问题，可以先浏览答案选项，查看所需的近似值的取值范围。在其他类型的问题中，如果读完题干没有思路，也可以试着浏览一下选项，以便更好地理解问题在问什么。

3. 不定项选择题

例题

Last year Kate spent between $\frac{1}{4}$ and $\frac{1}{3}$ of her gross income on her mortgage payments. If Kate spent $13,470 on her mortgage payments last year, which of the following could have been her gross income last year?

Indicate all such gross incomes.

☐ $40,200   ☐ $43,350   ☐ $47,256   ☐ $51,996   ☐ $53,808

如上题所示，这类题目要求考生从选项中选择一个或多个答案选项。这类问题可能会也可能不会指定要选出的选项数量。这些问题在答案选项旁边标有方框，而不是圆圈或椭圆形。请注意，正确答案也可以只有一个，并且多选、少选或错选均不给分。

识别方式：在实际考试中，不定项选择题的选项前面是方框，并且配有以下文字：Indicate all such statements/values…即选出所有符合条件的选项。

题干翻译

去年 Kate 花了她总收入的 $\frac{1}{4} \sim \frac{1}{3}$ 支付按揭贷款。如果 Kate 去年支付的按揭贷款为 13470 美元，以下哪些可能是她去年的总收入？

解题思路

设 Kate 去年的收入为 x,

根据题干可知, $\dfrac{1}{4} < \dfrac{13470}{x} < \dfrac{1}{3}$,

$40410 < x < 53880$,

因此选 \$43350, \$47256, \$51996 和 \$53808。

● 不定项选择题的做题建议：

（1）务必注意题干要求，看清楚是让你指出具体数值的答案选项还是所有适用的选项。在后一种情况下，一定要考虑所有的选项，以确定哪些是正确的。

（2）有些问题涉及选择可能的数值，此类题目建议确定最小可能值和最大可能值，这样可以有助于我们快速判断出可能的取值范围。

（3）部分题目应尝试去寻找数字模式和规律，以避免冗长的计算。

4. 数字填空题

In the xy-plane, what is the x-intercept of the line given by the equation $4x + 3y = 24$?

这类题目要么要求考生在一个答案框中输入整数或小数，要么要求考生在两个独立的框中输入分数的分子和分母。注意，分数不要求约分。使用电脑的鼠标和键盘输入数字答案。

题干翻译

在 xy 平面，直线方程 $4x + 3y = 24$ 的 x 轴截距是多少？

解题思路

题目要求计算 $4x + 3y = 24$ 的 x 轴截距。

令 $y = 0$，可求得 $x = 6$,

因此在答案框中输入数字 6。

● 数字填空题做题建议：

（1）务必仔细阅读题干要求，以确保填入的数字符合要求。有时在答案框的前面或后面会有标签来指示合适的答案类型。特别注意单位（如英尺或英里）和数量级（如百万或十亿等）。

（2）如果题目要求对答案进行四舍五入，请确保你四舍五入到要求的精度。例如，如果要将 46.7 四舍五入到最接近的整数,则需要输入数字47。如果题目没有四舍五入的要求,算出的数值是多少则填入多少。

（3）答案框中只能填入整数、小数和负号。

四、GRE 数学考试的注意事项

在 GRE 数学考试中,所有的题目都有统一的默认规则,这些规则说明出现在数学部分的最开始,重点如下:

（1）如果问题对应的选项前的符号为椭圆形,则正确答案只有 1 个;如果问题对应的选项前的符号为方形,则正确答案为 1 个或者多个。

（2）所有的数均为实数。GRE 数学中不考虑虚数。如果方程 $x^2 = -1$,则这个方程无解。

（3）如果没有特殊说明,所有的图形都默认在同一个平面内。

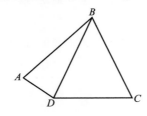

如上图所示,如果题目中没有强调 ABCD 不在同一平面内,我们应该把 ABCD 看成一个平面四边形。

（4）几何图形,如直线、圆、三角形和四边形,不一定按比例绘制。即我们无法在题目没有描述的情况下去根据图形判断线段长度和角度大小。

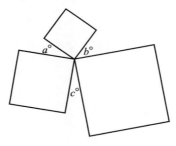

如上图,我们无法根据图形中显示的情况来确定 $\angle a$、$\angle b$ 和 $\angle c$ 的角度,以及它们之间的大小关系。但是,注意,图中如果显示的是直线,那么这条线一定是直线,并且直线上显示上的点的顺序,即为这些点的顺序,所有几何图形的相对位置是固定的。

如上图所示,虽然我们不知道 $\angle a$、$\angle b$ 和 $\angle c$ 的角度,以及它们之间的大小关系,但由于图里面 $\angle a$、$\angle b$ 和 $\angle c$ 有空隙,因此 $\angle a$、$\angle b$ 和 $\angle c$ 的角度均大于 0。

（5）坐标系,如直角坐标系、数轴是按比例绘制的,因此我们可以根据直角坐标系和数轴中点的位置读出该点的数值。

（6）图表,如柱状图、饼图、折线图等是按比例绘制的。

五、GRE 数学考试中的计算器

现在给大家介绍一下 GRE 数学界面的计算器。

当我们需要使用计算器协助计算的时候，点击界面的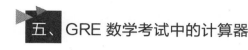键，系统自带的计算器就会跳出来。

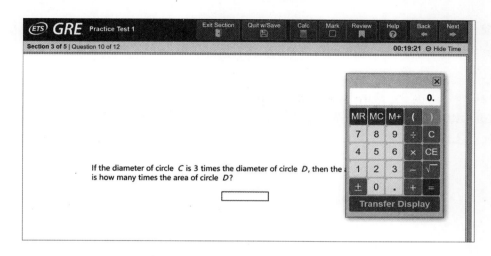

（1）GRE 数学涉及的复杂计算不多，并非每道题都需要用计算器。简单的计算建议大家自己算，这样更快：比如 $2^4 = 16$，$4 \times 70 = 280$，大家自己算就行。

（2）烦琐的较大数字的计算推荐用计算器，这样可以避免计算错误，也能节约运算时间。比如 19×59，这种计算并不是我们烂熟于心的乘法公式，建议用计算器。

（3）如果使用计算器，可以事先估计答案的大概范围，通过判断计算器的答案是否"在大致范围内"，可以帮助我们避免键入错误。比如 19×59，算出的值应该接近 $20 \times 60 = 1200$，如果你算出的结果与估算的值差距较大，要检查自己在使用计算器的时候是否有数字录入错误。特别是带小数点的计算，小数点的位置不要录入错误。

（4）如果数字填空题要求给出一个分数的答案，这个时候就不推荐用计算器算出这个值对应的小数了。当然，如果选项中给出的是分数，可以用计算器算出对应的数值，再看哪个选项对应的分数值与计算器的数值最为接近。

（5）计算器录入数字可以用鼠标点击计算器上的数值或用键盘录入，注意输入的数字以及相应的操作符号不要按错。

（6）关于计算器的按钮 Transfer Display 。这个键可以使我们把计算器中的数值直接转换到数值填空题的输入框中。但应用的时候要注意检查转移的数字是否符合问题的正确回答形式。例如，如果一个问题要求我们将计算的答案进行四舍五入或将计算的答案转换为百分比，请根据要求调整需要转移的数字。

（7）除了括号之外，屏幕上的计算器还有一个记忆栏和三个相应的按钮：记忆恢复键 MR、记忆清零键 MC 和记忆求和键 M+。这些按钮的功能和大多数计算器上的功能一样。（在目前的 GRE 数学考试中，基本不需要用到这些键，所以大家知道有这些键即可。）

（8）GRE 数学中的计算都默认是在实数范围内进行的操作，所以如果用计算器进行不在实数范围内的计算，计算器将会显示 Error。例如，除以零或取负数的平方根。如果输入 6 ÷ 0 =，则计算器将会显示 Error。同样，如果输入 1 ± √，则将显示 Error。要清除显示的结果，必须按清除按钮 C。

（9）计算器最多显示 8 位数字。即计算中涉及非常大的数字或者非常小的数字：如计算结果超过 8 位数的数字（比如计算 10,000,000 × 10 =），或者得出的计算结果为小于 0.00000001 的正数，计算器将无法正确显示或显示 0。如果 GRE 数学涉及此类数字，不推荐用计算器，题目应该是想考查一些小学奥数有关的解题技巧，咱们应该学会找规律，而不是用计算器硬算。

（10）注意 GRE 数学计算器只能帮忙做计算器上显示的运算操作，如加减乘除和开平方根。如果遇到这些情况以外的计算操作（如开三次方），是无法使用计算器的，我们需要另外了解一些此类题的考点和解题技巧。

第二篇

GRE 数学
考查的知识点

第一章 CHAPTER

数论

数论部分主要考查我们在小学学过的一些知识点。主要包括整数的性质和类型（如整除、因式分解、质数、余数、奇偶性）；分数和小数；指数和根；以及百分比、比例/比率等概念。大家千万不要因为是小学内容就轻视它，数论部分在 GRE 数学中的考查形式非常灵活，比较类似于小学奥数，是中国学生失分较多的部分之一。

数论部分主要考查6个知识点：
（1）整数
（2）分数和小数
（3）指数和根
（4）实数
（5）比例/比率
（6）百分比

1.1 整数

1.1.1 整数

基本词汇

integer 整数	add/plus 加	divisible 整除
positive integer 正整数	minus/subtract 减	sum 和
negative integer 负整数	multiply 乘	difference 差
zero 零	divide 除	

概念

整数是数字 1，2，3，…，以及它们的负数 –1，–2，–3，…和 0。因此，整数集是 {…，–3，–2，–1，0，1，2，3，…}。

正整数大于 0，负整数小于 0，0 既不是正整数也不是负整数。当整数相加、相减或相乘时，结果依旧是一个整数。整数的除法稍后再讨论。

考点

关于整数，我们需要掌握的性质如下：
- 两个正整数的乘积是一个正整数。
- 两个负整数的乘积是一个正整数。
- 一个正整数和一个负整数的乘积是负整数。

拓展性质

- 偶数个负整数的乘积是一个正整数。
- 奇数个负整数的乘积是一个负整数。

例题 01.

x, y, z, u, v, w, and t are seven nonzero integers. Which of the following values can be the number of these seven integers that are negative if $x \times y \times z = u \times y \times w \times t$?
Indicate <u>all</u> such values.

A. Three B. Four C. Five

题干翻译

x, y, z, u, v, w 和 t 是 7 个非零整数。如果 $x \times y \times z = u \times y \times w \times t$，以下哪个值可能是这 7 个整数中负数的个数？

解题思路

A 选项：如果负数为 3 个，为奇数，奇数 = 偶数 + 奇数，即左边和右边表达式中未知数是负数的个数只能是一边为奇数个，一边为偶数个，因此左边和右边的表达式必然一正一负，不可能相等，因此排除。

B 选项：如果负数为 4 个，比如左边 2 个负数，右边 2 个负数，这样左边和右边的表达式都是正数，可能相等。

C 选项：如果负数为 5 个，与 A 选项的分析同理，左边和右边的表达式必然一正一负，不可能相等。

答案 B

例题 02

r, s, and t are negative integers.

Quantity A	Quantity B
rst	$r^2 + s^2 + t^2$

A. Quantity A is greater.

B. Quantity B is greater.

C. The two quantities are equal.

D. The relationship cannot be determined from the information given.

题干翻译

r, s 和 t 是负整数。

解题思路

根据"奇数个负整数的乘积是一个负整数"可知，数量 A "3 个负整数 rst 的乘积"为负数。

根据非 0 数字的平方数大于 0 可知，数量 B "$r^2 + s^2 + t^2$ 3 个正数相加"，结果为正数。

即 $rst < 0 < r^2 + s^2 + t^2$。

故数量 B 更大。

答案 B

1.1.2 因数和倍数

> **基本词汇**
>
> factor/divisor 因数/因子
> multiple 倍数
>
> product 乘积
> X is divisible by Y. X 能被 Y 整除。

概念

当整数与整数相乘时，每个相乘的整数称为乘积的因数/因子。

例如，$2 \times 3 \times 10 = 60$，2，3 和 10 是 60 的因数。整数 4，15，5 和 12 也是 60 的因数，因为 $4 \times 15 = 5 \times 12 = 60$。所以，60 的正因数是 1，2，3，4，5，6，10，12，15，20，30 和 60。这些整数的负数也是 60 的因数，例如，$(-2) \times (-30) = 60$，-2 与 -30 也是 60 的因数。同时，60 是其每个因数的倍数，60 可以被它的每个因数整除。

考点 1　因数的性质

如果题目中说某个数 A 是另一个数 B 的因数，则在暗示 A 这个数所有的因数完全被 B 包含。

此外，还需要注意两个隐蔽的性质：

- 1 是任何整数的因数。
- 0 是任何整数的倍数。

考点 2　数的整除特征

GRE 数学要求考生可以快速判断 2，3，4，5，8，9，25，125 的倍数，这些倍数的判断方法如下：

1. 只看个位数——2，5 的倍数

（1）2 的倍数的特征：个位数字是 0，2，4，6，8 的整数。

（2）5 的倍数的特征：个位数字是 0 或 5。

2. 只看末两位数——4，25 的倍数

（3）4 的倍数的特征：末两位是 4 的倍数。

（4）25 的倍数的特征：末两位是 25 的倍数。

3. 看末三位数——8，125 的倍数

（5）8 的倍数的特征：末三位是 8 的倍数。

（6）125 的倍数的特征：末三位是 125 的倍数。

4. 看各个数位上的数字相加的和——3,9 的倍数

（7）3 的倍数的特征：各个数位上的数字之和是 3 的倍数。

（8）9 的倍数的特征：各个数位上的数字之和是 9 的倍数。

例题 01.

If even integer x is divisible by positive odd integer y, then x^2 must be divisible by

A. $y+4$ B. $3y$ C. $4y$ D. $4y+1$ E. $4(y+1)$

题干翻译

如果偶数 x 可以被正奇整数 y 整除，那么 x^2 一定可以被什么整除？

解题思路

根据题干条件"x 是偶数"可知，x 是 2 的倍数，

根据题干条件"x 可以被 y 这个奇数整除"可知，x 是 y 的倍数，

因此 x 是 2 且是 y 的倍数，即 x 是 $2y$ 的倍数，

x^2 是 $4y^2$ 的倍数，即 x^2 包含 $4y^2$ 的所有因子，那么 x^2 包含选项 C 的 $4y$。

答案　C

例题 02.

Let $y=105n$, where n is a positive integer. If y is the square of an integer and y is divisible by 30, what is the least possible value of n?

A. 105 B. 210 C. 420 D. 1,470 E. 3,150

题干翻译

如果 $y=105n$，n 为正整数。如果 y 是一个整数的平方，并且 y 可以被 30 整除，那么 n 的最小可能值是多少？

解题思路

根据题干条件可知，$y=105n=3\times5\times7n$，

根据题干条件"y 可以被 30 整除"可知，y 需要包含 30 所有的质因数（2, 3, 5），

又根据题干条件"y 是一个平方数"可知，y 需要包含 2, 3, 5, 7 的平方，

因此，y 的最小可能值 $=2^2\times3^2\times5^2\times7^2$，

n 的最小可能值 $=\dfrac{2^2\times3^2\times5^2\times7^2}{105}=420$。

答案　C

例题 03

The 4-digit positive integer $8R5T$, where R is the hundreds digit and T is the units digit, is divisible by 9. Which of the following could be the value of $R + T$?

Indicate <u>all</u> such values.

A. 2   B. 3   C. 5   D. 11   E. 14

题干翻译

$8R5T$ 是一个可以被 9 整除的 4 位数,其中 R 是百位数位对应的数值,T 是个位数位对应的数值。以下哪一项可能是 $R + T$ 的值?

解题思路

能够被 9 整除的数字特点是:各个数位上的数字之和是 9 的倍数,

根据题干条件 "$8R5T$ 可以被 9 整除" 可知,$8 + R + 5 + T$ 为 9 的倍数,

即 $13 + R + T$ 为 9 的倍数,

当 $R + T = 5$,$13 + R + T = 18$,为 9 的倍数,正确;

当 $R + T = 14$,$13 + R + T = 27$,为 9 的倍数,正确。

答案 C E

练习

1. If $n = 3^8 - 2^8$, which of the following is NOT a factor of n?

   A. 97   B. 65   C. 35   D. 13   E. 5

2. A certain restaurant offers each customer a combination dinner consisting of a choice of any entrée, a choice of any beverage, and a choice of any dessert. The number of different combination dinners that are possible is 90. Which of the following CANNOT be the number of desserts available to be chosen for a combination dinner?

   A. 2   B. 3   C. 4   D. 5   E. 6

3. There is a six-digit integer: $639,k70$. Which of the following cannot be the factor of the integer?

   A. 2   B. 3   C. 4   D. 5   E. 7

答案及解析

1. If $n = 3^8 - 2^8$, which of the following is NOT a factor of n?

A. 97   B. 65   C. 35   D. 13   E. 5

题干翻译

如果 $n = 3^8 - 2^8$，以下哪一项不是 n 的因数？

解题思路

如果选项是 n 的因数，那么 n 肯定包含这个选项；如果选项不是 n 的因数，那么 n 不包含这个选项。因此题目本质问的是 n 不包含哪个选项。

利用平方差公式：$a^2 - b^2 = (a-b)(a+b)$，可知：

$n = 3^8 - 2^8 = (3^4 - 2^4)(3^4 + 2^4)$,
$\quad = (81 - 16)(81 + 16)$,
$\quad = 65 \times 97 = 5 \times 13 \times 97$,

因此 n 包含 5，13，65，97，不包含 35。

答案　C

2. A certain restaurant offers each customer a combination dinner consisting of a choice of any entrée, a choice of any beverage, and a choice of any dessert. The number of different combination dinners that are possible is 90. Which of the following CANNOT be the number of desserts available to be chosen for a combination dinner?

A. 2   B. 3   C. 4   D. 5   E. 6

题干翻译

某家餐厅为每位顾客提供组合晚餐，包括任意一种主菜、任意一种饮料和任意一种甜点。可能的不同组合晚餐的数量是 90。下列哪一项不能成为组合晚餐可供选择的甜点数量？

解题思路

组合晚餐的数量 = 主菜的数量 × 饮料的数量 × 甜点的数量，

即 90 = 主菜的数量 × 饮料的数量 × 甜点的数量，甜点数量为 90 的因数，

因此本题的本质是在问以下哪个选项不是 90 的因数，

90 无法被 4 整除，即 4 不是 90 的因数。

答案　C

3. There is a six-digit integer：639, k70. Which of the following cannot be the factor of the integer?

A. 2   B. 3   C. 4   D. 5   E. 7

题干翻译

有一个六位数整数：639k70。以下哪一项不可能是这个整数的因数？

解题思路

根据数的整除特征可知，是不是 4 的倍数，看的是末两位。

639k70 的末两位数为 70，70 不能被 4 整除，因此 4 不是 639k70 的因数。

答案　C

1.1.3　最大公因数和最小公倍数

基本词汇

the greatest common divisor/factor 最大公因数　　the least common multiple 最小公倍数

概念

最大公因数，也称为最大公约数，最大公因子，指多个整数共有的因数中最大的那一个。

最小公倍数，指的是多个整数除了 0 以外的共同倍数里最小的那一个。

考点

最大公因数、最小公倍数主要考查如何求值。

最大公因数和最小公倍数的求值方法如下：

(1) 对所列出的数字进行分解质因数。

(2) 取共同质因数的最小指数形式，相乘得到最大公因数。

(3) 取所有不同质因数的最大指数形式，相乘得到最小公倍数。

我们来举个例子：

求 90，196，200 的最大公因数和最小公倍数。

步骤 1：先对这些数字进行分解质因数，

$$90 = 2 \times 3^2 \times 5,$$
$$196 = 2^2 \times 7^2,$$
$$200 = 2^3 \times 5^2。$$

步骤 2：最大公因数取共同质因数的最小指数形式相乘，所以最大公因数 = 2。

最小公倍数取所有不同质因数的最大指数形式相乘，所以最小公倍数 = $2^3 \times 3^2 \times 5^2 \times 7^2 = 88200$。

那么，考试的时候又是如何考查的呢？下面我们来看几道例题。

If $x = k^2n^3p$ and $y = k^3np^2$, where k, n, and p are different prime numbers, which of the following is the least common multiple of x and y?

A. k^2n^3p　　B. $k^3n^3p^2$　　C. $k^3n^3p^3$　　D. $k^5n^4p^3$　　E. $k^6n^3p^2$

题干翻译

如果 $x = k^2 n^3 p$，$y = k^3 n p^2$，其中 k，n，p 是不同的质数，以下哪一项为 x 和 y 的最小公倍数？

解题思路

计算 x 和 y 的最小公倍数，根据上述方法，我们知道：

① 对所列出的数字进行分解质因数（由于题干中已经给出了这步答案，咱们可以直接用）；

② 取所有不同质因数的最大指数形式，相乘得到最小公倍数（即 $k^3 n^3 p^2$）。

答案　B

Let $m = (2^3)(3^2)(5)(7^2)$ and $p = (2^2)(3^5)(5^4)(11)$. What is the greatest common divisor of m and p?

A. $(2)(3)(5)$　　　　　　　B. $(2^2)(3^2)(5)$　　　　　　　C. $(2)(3)(5)(7)(11)$

D. $(2^2)(3^2)(5)(7)(11)$　　　E. $(2^3)(3^5)(5^4)(7^2)(11)$

题干翻译

$m = (2^3)(3^2)(5)(7^2)$，$p = (2^2)(3^5)(5^4)(11)$。m 和 p 的最大公因数是多少？

解题思路

计算 m 和 p 的最大公因数，根据上述方法，我们知道：

① 对所列出的数字进行分解质因数（由于题干中已经给出了这步答案，咱们可以直接用）；

② 取共同质因数的最小指数形式，相乘得到最大公因数，即 $(2^2)(3^2)(5)$。

答案　B

1. d is the greatest common divisor of 36 and 60, and m is the least common multiple of 36 and 60.

Quantity A	Quantity B
$\dfrac{36}{d}$	$\dfrac{m}{60}$

A. Quantity A is greater.

B. Quantity B is greater.

C. The two quantities are equal.

D. The relationship cannot be determined from the information given.

2. The greatest common factor of the positive integers a and b is 3, and $5a = 7b$. What is the least common multiple of a and b?

   A. 35   B. 70   C. 105   D. 210   E. 315

3. The greatest common divisor of two positive integers k and n is 5, and the least common multiple of k and n is 30. If $k = 10$, what is the value of n?

   A. 5   B. 10   C. 15   D. 20   E. 30

答案及解析

1. d is the greatest common divisor of 36 and 60, and m is the least common multiple of 36 and 60.

Quantity A	Quantity B
$\dfrac{36}{d}$	$\dfrac{m}{60}$

   A. Quantity A is greater.
   B. Quantity B is greater.
   C. The two quantities are equal.
   D. The relationship cannot be determined from the information given.

   题干翻译
   d 是36和60的最大公因数，m 是36和60的最小公倍数。

   解题思路
   计算36和60的最大公因数和最小公倍数。
   根据上述方法，我们知道：
   ① 对所列出的数字进行分解质因数 $36 = 2^2 \times 3^2$，$60 = 2^2 \times 3 \times 5$，
   ② 取共同质因数的最小指数形式，相乘得到最大公因数 $d = 2^2 \times 3^1 = 12$，
   ③ 取所有不同质因数的最小指数形式，相乘得到最小公倍数 $m = 2^2 \times 3^2 \times 5^1 = 180$，
   数量 A $= \dfrac{36}{d} = \dfrac{36}{12} = 3$，数量 B $= \dfrac{m}{60} = \dfrac{180}{60} = 3$，即数量 A = 数量 B。

   答案 C

2. The greatest common factor of the positive integers a and b is 3, and $5a = 7b$. What is the least common multiple of a and b?

   A. 35   B. 70   C. 105   D. 210   E. 315

   题干翻译
   正整数 a 和 b 的最大公因数为3，并且 $5a = 7b$。a 和 b 的最小公倍数是多少？

解题思路

根据题干条件"a 和 b 的最大公因数是 3"可知，a 和 b 都包含 3。

根据题干条件 $5a = 7b$ 可知，a 包含 7，b 包含 5。

因此 a 和 b 的最小公倍数 $= 3 \times 7 \times 5 = 105$。

答案 C

3. The greatest common divisor of two positive integers k and n is 5, and the least common multiple of k and n is 30. If $k = 10$, what is the value of n?

A. 5　　　　　B. 10　　　　　C. 15　　　　　D. 20　　　　　E. 30

题干翻译

k 和 n 这两个正整数的最大公约数为 5，k 和 n 的最小公倍数为 30。如果 $k = 10$，那么 n 是多少？

解题思路

根据题干条件"$k = 10$，且 k 和 n 的最大公约数为 5"可知，n 中包含 5，而且不包含 2；

根据题干条件"k 和 n 的最小公倍数为 $30 = 2 \times 3 \times 5$"可知，k 或 n 中还包含质因数 3。

显然 $k = 10$ 不包含 3，说明 3 是被 n 包含的。

因此 $n = 3 \times 5 = 15$。

答案 C

1.1.4 商和余数

基本词汇

| quotient 商 | A is divided by B. A 除以 B。（不知道能不能被整除） |
| remainder 余数 | A is divisible by B. A 除以 B，且可以被整除。 |

概念

当两个整数相除且不能整除的时候，结果可以写成一个分数或小数（这两者都将在后面讨论），或者结果可以看作一个带有余数的商。

正整数 X ÷ 正整数 Y，整除得到的值叫作商；不能整除的部分，叫作余数。

被除数 ÷ 除数 = 商……余数

例如：$8 \div 5 = 1 \cdots\cdots 3$，其中商就是 1，余数是 3。

考点1　余数的大小关系

首先，余数必须是 ≥ 0 的整数。

正整数 X ÷ 正整数 Y 刚好可以整除的话，余数为 0。

其次，余数必须小于除数 Y。

综上，$0 \leq$ 余数 $<$ 除数。

考点 2 余数运算和除法运算

除法运算：$8 \div 5 = 1.6$，

余数运算：$8 \div 5 = 1 \cdots\cdots 3$，其中 3 是余数。

> **注意**
>
> （1）除法运算是可以进行约分的，但因为余数大小受到除数大小的约束，因此余数运算不能进行约分。
>
> 除法运算：$16 \div 10 = 8 \div 5 = 1.6$，
>
> 余数运算：$16 \div 10 = 1 \cdots\cdots 6$。
>
> （2）在除法运算中，得到的数值的小数部分 × 除数 = 余数。
>
> $8 \div 5 = 1.6$，相当于 $8 = 1.6 \times 5 = 1 \times 5 + 0.6 \times 5$，
>
> 其中 1×5 表示可以整除的部分，而 0.6×5 表示不能整除的部分，
>
> 由于余数指的就是不能够整除的部分，
>
> 所以 $0.6 \times 5 = 3$（余数）。

考点 3 整除的公式表达

When positive integer n is divided by 8, the remainder is 5. What is the remainder when n is divided by 4?

题干翻译

当正整数 n 除以 8 时，余数是 5。n 除以 4 的余数是多少?

解题思路

根据题干条件可知，$n = 8a + 5$（a 为变量）。

虽然有 a 这个变量，但变量 a 前的系数 8 可以被 4 整除，因此 $8a$ 这个部分除以 4 不会产生余数，$8a + 5$ 除以 4 的余数由 5 除以 4 的余数决定，5 除以 4 余 1，因此 n 除以 4 的余数是 1。

答案 1

> 当我们遇到含有变量的表达式，问该表达式除以除数的余数是多少时，如果变量前的系数能够被除数整除，那么变量的变化不影响余数，余数部分由常数除以除数来决定。

例题 02.

When positive integer n is divided by 8, the remainder is 5. What is the remainder when n is divided by 6?

题干翻译

当正整数 n 除以 8 时，余数是 5。n 除以 6 的余数是多少?

解题思路

根据题干条件可知，$n = 8a + 5$（a 为变量）。

此时，变量 a 前的系数 8 没法被 6 整除，因此变量 a 的变化会影响余数。

这种情况下，我们通过试数，找出余数的循环即可。

当 $a=1$ 时，$n=13$，$n\div 6$ 余 1；

当 $a=2$ 时，$n=21$，$n\div 6$ 余 3；

当 $a=3$ 时，$n=29$，$n\div 6$ 余 5；

当 $a=4$ 时，$n=37$，$n\div 6$ 余 1；

当 $a=5$ 时，$n=45$，$n\div 6$ 余 3；

……

这样我们就可以看出 $n=8a+5$ 的余数为 1，3，5 这三种情况一直循环。

答案 1，3，5

总结 当我们遇到含有变量的表达式，问该表达式除以除数的余数是多少时，如果变量前的系数不能够被除数整除，那么变量的变化会影响余数。在这种情况下，我们可以通过以下方式来确定余数的可能值：往变量里代入具体数值，算出表达式除以除数后的余数，找出余数的循环模式。

例题 03

When positive integer n is divided by 12, the remainder is 5. When positive integer n is divided by 18, the remainder is 11. What is the remainder if positive integer n is divided by 36?

题干翻译

当正整数 n 除以 12 时，余数是 5。当正整数 n 除以 18 时，余数是 11。正整数 n 除以 36 的余数是多少？

解题思路

根据题干条件可知，$n=12a+5$（a 为变量），且 $n=18b+11$（b 为变量）。

如果我们遇到一个未知数给出了两个表达式的情况，建议将表达式合并，再根据题干的要求进行进一步的变形。

两个表达式的合并规则：

① 新的表达式未知数前的系数取原本两个表达式系数的最小公倍数，本题中未知数前的系数为 12 和 18，12 和 18 的最小公倍数为 36。

② 新的表达式的常数项，是通过在原本两个表达式中试数，找出的两个表达式中共有的数字。

比如，满足 $12a+5$ 的数字有 5，17，29，…，满足 $18b+11$ 的数字有 11，29，47，…。我们可以发现这两个式子中都出现了数字 29。

因此，我们可以推出 $n=36c+29$（c 为变量）。

题干问 n 除以 36 的余数，我们发现 $36c+29$ 这个表达式中未知数前的系数可以被 36 整除，因此变量的变化不影响余数，因此 $36c+29$ 除以 36 的余数为 29。

答案 29

> 如果我们遇到一个未知数给了2个表达式的情况,我们的建议是将表达式合并,再根据题干的要求进行进一步的变形。
> 2个表达式的合并规则:
> (1) 新的表达式未知数前的系数取原本两个表达式系数的最小公倍数。
> (2) 新的表达式的常数项,是通过在原本两个表达式中试数,找出的两个表达式中共有的数字。

When positive integer A is divided by positive B, the remainder is 6. If A is less than 8, what is the value of A?

A. 5   B. 6   C. 7   D. 8   E. 9

题干翻译

正整数 A ÷ 正整数 B 的余数是 6。已知 $A<8$,问 A 的值是多少?

解题思路

根据余数的取值范围:0≤余数<除数,

因此 $A>6$,且根据题干条件 $A<8$ 可知,$A=7$。

答案   C

When positive integer x is divided by positive integer y, the remainder is 6. If $\frac{x}{y} = 56.12$, what is the value of y?

A. 64   B. 50   C. 32   D. 16   E. 8

题干翻译

正整数 x ÷ 正整数 y 的余数是 6,且 $\frac{x}{y}=56.12$,y 的值是多少?

解题思路

在除法运算中,得到的数值的小数部分×除数=余数。

所以,$0.12 \times y = 6$,$y = 6 \div 0.12 = 50$。

答案   B

例题 06.

The remainder is 19 when a positive integer n is divided by 42, what is the remainder when n is divided by 7?

题干翻译

正整数 n 除以 42 的余数是 19,n 除以 7 的余数是多少?

解题思路

根据题干条件"n 除以 42 的余数是 19"可知，$n = 42a + 19$（a 为变量）。

由于变量 a 前的系数可以被 7 整除，因此 a 的变化不影响 n 除以 7 的余数。

n 除以 7 的余数由常数部分除以 7 产生：$19 \div 7 = 2 \cdots\cdots 5$，

即 $n \div 7$ 的余数为 5。

答案 5

a is a positive integer.

Quantity A	Quantity B
x is the remainder when $15a$ is divided by 6.	2

A. Quantity A is greater.

B. Quantity B is greater.

C. The two quantities are equal.

D. The relationship cannot be determined from the information given.

题干翻译

a 是一个正整数。

解题思路

本题让我们比较的是"$15a$ 除以 6 的余数"与 2 的大小关系。

由于变量 a 前面的系数 15 无法被 6 整除，因此 $15a$ 除以 6 的余数我们应该通过试数找规律得出：

当 $a = 1$，$15a = 15$，$15 \div 6$ 的余数 $3 > 2$；

当 $a = 2$，$15a = 30$，$30 \div 6$ 的余数 $0 < 2$。

因此，数量 A 与数量 B 的大小关系无法确定。

答案 D

When the positive integer n is divided by 4, the remainder is 3; when n is divided by 3, the remainder is 2.

Quantity A	Quantity B
The least possible value of n	12

A. Quantity A is greater.

B. Quantity B is greater.

C. The two quantities are equal.

D. The relationship cannot be determined from the information given.

题干翻译

当正整数 n 除以 4 时，余数是 3；当 n 除以 3 时，余数是 2。

解题思路

前面说过，如果遇到一个未知数给出了两个表达式的情况，建议将表达式合并，再根据题干的要求进行进一步的变形。

两个表达式的合并规则：
① 新的表达式未知数前的系数取原本两个表达式系数的最小公倍数。
② 新的表达式的常数项，是通过在原本两个表达式中试数，找出两个表达式中共有的数字。

根据题干条件可知，$n=4a+3$，$n=3b+2$。

未知数 c 前的系数 12 取的是原本两个表达式前 4 和 3 的最小公倍数 12。

常数由试数找出重复的数得出，$4a+3$ 对应的数字有 3，7，11，15，…，$3b+2$ 对应的数字有 2，5，8，11，14，…，两个表达式中共有的数字为 11。

因此，满足要求的 $n=12c+11$，

数量 A 为 n 的最小可能值，n 为正整数，且要符合 $12c+11$。

故 n 的最小可能值是 11，<12。

数量 B 更大。

答案 B

练 习

1. When positive integer m is divided by 6, the remainder is 4. When positive integer p is divided by 6, the remainder is 5. What is the remainder when the product mp is divided by 6?

   A. 1   B. 2   C. 3   D. 4   E. 5

2. When positive integer n is divided by 5, the remainder is 2; when n is divided by 6, the remainder is 3. If $0 < n < 100$, what is the number of n?

   A. 1   B. 2   C. 3   D. 4   E. 5

3. Prime number n is less than 100. When n is divided by 5, the remainder is 2; when n is divided by 7, the remainder is 6. What is the remainder when n is divided by 8?

答案及解析

1. When positive integer m is divided by 6, the remainder is 4. When positive integer p is divided by 6, the remainder is 5. What is the remainder when the product mp is divided by 6?

A. 1　　　　B. 2　　　　C. 3　　　　D. 4　　　　E. 5

题干翻译

当正整数 m 除以6时，余数是4。当正整数 p 除以6时，余数是5。乘积 mp 除以6的余数是多少？

解题思路

根据题干条件可知：

$m = 6a + 4$（a 为变量），

$p = 6b + 5$（b 为变量），

因此 $mp = (6a+4)(6b+5) = 36ab + 30a + 24b + 20$。

由于变量 ab，a，b 前的系数都可以被 6 整除，因此 mp 除以 6 的余数由常数除以 6 来决定：$20 \div 6 = 3 \cdots\cdots 2$。

答案　B

2. When positive integer n is divided by 5, the remainder is 2; when n is divided by 6, the remainder is 3. If $0 < n < 100$, what is the number of n?

A. 1　　　　B. 2　　　　C. 3　　　　D. 4　　　　E. 5

题干翻译

正整数 n 除以5，余数为2；当 n 除以6时，余数是3。如果 $0<n<100$，n 的个数是多少？

解题思路

根据题干条件可知：$n = 5a + 2$，$n = 6b + 3$。

因此，$n = 30c + 27$。

由于 $0 < n < 100$，

因此满足要求的数为 27，57，87 这三个数字。

答案　C

3. Prime number n is less than 100. When n is divided by 5, the remainder is 2; when n is divided by 7, the remainder is 6. What is the remainder when n is divided by 8?

题干翻译

质数 n 小于100。当 n 除以5时，余数是2；当 n 除以7时，余数是6。n 除以8的余数是多少？

解题思路

根据题干条件可知，$n = 5a + 2$，$n = 7b + 6$。

因此，$n = 35c + 27$。

由于 n 是 <100 的质数，所以，n 为 97。$97 \div 8 = 12 \cdots\cdots 1$。

答案　1

1.1.5　奇偶性

基本词汇

odd number 奇数　　　　　　　　　　　　　　even number 偶数

概念

在整数中，是 2 的倍数的数叫作偶数，不是 2 的倍数的数叫作奇数。我们通常用 $2n$ 来表示偶数，用 $2n+1$ 来表示奇数。

奇数、偶数的前提是整数，一个整数无论是正数、负数，还是零，都具备奇偶性（0 是偶数）。比如 1 是奇数，-1 也是奇数。正负不影响奇偶性。

考点 1　加减乘除运算对奇偶性判断的影响

1. 加减运算

$a + b = \text{odd}$，说明 a 和 b 奇偶性相反，一奇一偶，但是谁奇谁偶无法确定。
$a + b = \text{even}$，说明 a 和 b 奇偶性相同，可能同为奇数，也可能同为偶数。
减法同理。
2 个数字的加减运算的规则：一奇一偶相加减结果为奇，同奇同偶相加减结果为偶。
拓展规则：
$a + b + \cdots + n = \text{odd}$，如果多个数相加，结果为奇数，说明里面有奇数个奇数。
$a + b + \cdots + n = \text{even}$，如果多个数相加，结果为偶数，说明里面有偶数个奇数。

2. 乘法运算

$a \times b \times c \times d = \text{odd}$，说明乘项 a，b，c，d 都是奇数。
$a \times b \times c \times d = \text{even}$，说明乘项中至少有 1 个偶数。
乘法运算规则：乘积为奇，说明乘项都是奇数；乘积为偶，说明乘项中至少有一个偶数。

3. 除法运算

$\dfrac{a}{b} = \text{even}$，如果遇到除法运算，建议转换成乘法形式：$a = b \times \text{even}$。

因为乘项中只要有偶数，乘积就一定是偶数。所以 a 一定是偶数。同时，因为奇数、偶数都是整数，所以，如果 $\dfrac{a}{b} = $ 整数，意味着 a 可以整除 b，a 是 b 的倍数。

考点 2　奇偶性的性质

1. 指数不影响奇偶性

如果 a 与 n 都是正整数，那么 a 和 a^n 的奇偶性完全一致。
$a^n = a \times a \times \cdots \times a$（$n$ 个 a 相乘），
如果 a 是奇数，乘项都是奇数，因此 a^n 还是奇数，
如果 a 是偶数，乘项中有偶数，那么 a^n 还是偶数。
因此，a^n 的奇偶性由底数 a 决定，与指数 n 无关（注意 n 必须是正整数）。

2. 相邻整数奇偶性相反，一奇一偶，乘积一定是偶数

由于相邻数字是奇偶间错分布的，即相邻的两个整数 n 和 $n+1$ 肯定是一个奇数一个偶数，虽然 n 与 $n+1$ 谁奇谁偶无法确定，但是它们的乘积 $n(n+1)$ 一定是偶数。

例题 01.

If x and y are positive integers and $x + y = 8x + 22$, which of the following must be true?

A. x is even.　　B. xy is odd.　　C. $x - y$ is odd.
D. $x(y+1)$ is even.　　E. x and y are both odd.

题干翻译

如果 x 和 y 是正整数，且 $x + y = 8x + 22$，以下哪一个表达一定是正确的？

解题思路

题干条件是：$x + y = 8x + 22$，
由于 $8x$ 与 22 都是偶数，我们可以得出 $x + y$ 是偶数，
因此，x 与 y 要么同为奇数要么同为偶数。
A 选项：x 为偶数。
因为 x 可以是奇数也可以是偶数，所以 A 选项错误。
B 选项：xy 为奇数。
当 x 与 y 都是偶数的时候，xy 为偶数，B 选项错误。
C 选项：$x - y$ 为奇数。
无论 x 与 y 同为奇数还是同为偶数，$x - y$ 都是偶数，C 选项错误。
D 选项：$x(y+1)$ 为偶数。
因为 x 与 y 奇偶性相同，y 与 $y+1$ 奇偶性相反（相邻数字奇偶性相反），所以可以得出 x 与 $y+1$ 奇偶性相反，即 x 与 $y+1$ 一个为奇数一个为偶数。根据乘法性质，乘项中只要有偶数，乘积结果为偶数，D 选项正确。
E 选项：x 与 y 均为奇数。
x 与 y 可以同为偶数，E 选项错误。

答案　D

If x and y are integers, $2y - x = x^2 - y^2$, which of the following must be an even integer? Indicate all such values.

A. x B. y C. $2x + y$ D. $y^2 + x$ E. $y^2 + y$

题干翻译

如果 x 和 y 是整数,且 $2y - x = x^2 - y^2$,下列哪一项一定是偶数?

解题思路

方法 1:

对于等式涉及多个未知数的情况,我们一般把同样的未知数写在同一侧。

因此,$2y - x = x^2 - y^2$ 可以写成 $2y + y^2 = x^2 + x = x(x+1)$。

因为 x 与 $x + 1$ 为相邻整数,相邻整数奇偶性相反,乘积一定是偶数,所以 $x(x+1)$ 肯定为偶数。因此 $2y + y^2$ 也为偶数。

根据"两个数字相加结果为偶数,这两个数字一定同奇同偶"的原则,且 $2y$ 为偶数,可知 y^2 为偶数。

由于指数不影响奇偶性,我们可以进一步推出 y 为偶数。

因此根据题干条件,x 的奇偶性无法判断,y 为偶数。

方法 2:

由于指数不影响奇偶性,因此 x 与 x^2 奇偶性一致。

根据 $2y - x = x^2 - y^2$ 这个等式可知,等式左侧与右侧奇偶性一致。

由以上两个条件我们可以推出 $2y$ 与 y^2 奇偶性一致,由于 $2y$ 为偶数,因此我们可以推出 y^2 为偶数。

由于指数不影响奇偶性,我们可以进一步推出 y 为偶数。

综上,x 的奇偶性无法判断,y 为偶数。

现在我们来依次看每一个选项:

A 选项:x,奇偶性不确定。

B 选项:y 为偶数,正确。

C 选项:$2x + y$,$2x$ 为偶数,y 为偶数,因此 C 选项为两个偶数相加,结果是偶数,正确。

D 选项:$y^2 + x$,由于 x 的奇偶性不确定,因此 $y^2 + x$ 整体奇偶性不确定。

E 选项:$y^2 + y$,为两个偶数相加,结果为偶数,正确。

答案 BCE

Nine integers are to be selected so that their product will be even.

Quantity A	Quantity B
The least possible number of the nine integers that can be even	2

A. Quantity A is greater.
B. Quantity B is greater.
C. The two quantities are equal.
D. The relationship cannot be determined from the information given.

题干翻译

选出 9 个整数，使它们的乘积为偶数。

解题思路

注意，the least possible number 指的是最小可能的个数，因此数量 A 指的是 9 个整数中偶数最小可能的个数。

因此，本题要我们比较的是 9 个整数中偶数最小可能的个数和数字 2 的大小关系。

根据奇偶性的乘法运算可知，乘积为偶数，说明乘项中至少有 1 个偶数。

在这 9 个整数的乘积为偶数的情况下，偶数最小可能的个数为 1。因此，数量 A 为 1。

1 < 2，故数量 B 更大。

答案　B

If a and b are positive integers such that $a - b$ and $\dfrac{a}{b}$ are both even integers, which of the following must be an odd integer?

A. $\dfrac{a}{2}$ B. $\dfrac{b}{2}$ C. $\dfrac{a+b}{2}$ D. $\dfrac{a+2}{2}$ E. $\dfrac{b+2}{2}$

题干翻译

如果 a 和 b 是正整数，使得 $a-b$ 和 $\dfrac{a}{b}$ 都是偶数，那么下列哪一项一定是奇数？

解题思路

根据题干条件 $a - b =$ even 可知，a 和 b 的奇偶性相同，因为同奇同偶相加减结果是偶数。

根据题干条件 $\dfrac{a}{b} =$ even 可知，$a = b \times$ even，因此 a 是偶数。

由此我们可以进一步推出 b 是偶数，$a =$ 偶数 $b \times$ even。

偶数是 2 的倍数，因此 a 是两个 2 的倍数的乘积，即 a 是 4 的倍数。

A 选项：$\dfrac{a}{2}$，因为 a 是 4 的倍数，因此 $\dfrac{a}{2}$ 是 2 的倍数，即 $\dfrac{a}{2}$ 为偶数，A 选项错误。

B 选项：$\dfrac{b}{2}$，由题干条件我们只知道 b 是偶数，而 $\dfrac{偶数}{2}$ 的奇偶性是不确定的。比如，$b = 2$ 的时候，$\dfrac{b}{2} = 1$ 为奇数；$b = 4$ 的时候，$\dfrac{b}{2} = 2$ 为偶数。因此 B 选项错误。

C 选项：$\frac{a+b}{2}$，根据前面的判断，我们知道$\frac{a}{2}$为偶数，$\frac{b}{2}$奇偶性不确定，因此$\frac{a+b}{2}$的奇偶性不确定，C 选项错误。

D 选项：$\frac{a+2}{2}$，因为$\frac{a}{2}$为偶数，$\frac{2}{2}$为奇数，因此$\frac{a+2}{2}$ = 偶数 + 奇数 = 奇数，正确。

E 选项：$\frac{b+2}{2}$，因为$\frac{b}{2}$奇偶性不确定，因此$\frac{b+2}{2}$的奇偶性不确定，E 选项错误。

注意：涉及对于表达式$\frac{x}{2}$奇偶性的判断，隐含的要求是让我们判断 x 是否是 4 的倍数。如果 x 是 4 的倍数，那么$\frac{x}{2}$为偶数；如果无法确定 x 是 4 的倍数，那么$\frac{x}{2}$的奇偶性不确定。

答案　D

练 习

1. If r, s, t, and u are positive integers such that $(r+s)(t+u)$ is an odd integer and $(r+s+t)(s+t+u)$ is an odd integer, which of the following statements must be true?
   A. s and t are both even.
   B. s and t are both odd.
   C. s is even and t is odd.
   D. s is odd and t is even.
   E. s is even if t is odd, and s is odd if t is even.

2. If $x = 2y + 1$, and $y = 2w$, where w, x, and y are integers, which of the following must be an odd integer?
   A. $xy + w$ B. $xy + w + 1$ C. $(x+y)w$ D. $wy + x$ E. $wx + y$

3. If k and n are positive integers such that $k + n$ is an odd integer, which of the following must be an odd integer?
   A. kn B. $kn + n$ C. $kn + n^2$ D. $(k-1)^2 + n^2$ E. $k^2 + n$

4. If x and y are both odd integers, which of the following must be a multiple of 4?
   A. $x^2 y$ B. $x^2(y+1)$ C. $x(y+1)$ D. $(x-1)y^2$ E. $(x+1)(y-1)$

5. If both the sum and the product of the four positive integers a, b, c, and d are even, what is the greatest number of these integers that could be odd?
   A. None   B. One   C. Two   D. Three   E. Four

答案及解析

1. If r, s, t, and u are positive integers such that $(r+s)(t+u)$ is an odd integer and $(r+s+t)(s+t+u)$ is an odd integer, which of the following statements must be true?

A. s and t are both even.
B. s and t are both odd.
C. s is even and t is odd.
D. s is odd and t is even.
E. s is even if t is odd, and s is odd if t is even.

题干翻译

如果 r, s, t 和 u 是正整数，$(r+s)(t+u)$ 是奇数，$(r+s+t)(s+t+u)$ 是奇数，下列哪个陈述一定是正确的？

解题思路

$(r+s)(t+u) = $ odd，根据乘法性质，乘积结果为奇数，说明乘项都是奇数，

即 $r+s = $ odd 且 $t+u = $ odd。

根据加法性质，两个整数相加，结果为奇数，说明两个整数奇偶性相反，一奇一偶。

即 r 和 s 的奇偶性不同，t 和 u 的奇偶性不同。

$(r+s+t)(s+t+u) = $ odd，根据乘法性质，乘积结果为奇数，说明乘项都是奇数，

即 $r+s+t = $ odd 且 $s+t+u = $ odd。

已知 $r+s = $ odd，说明 t 是 even，因此 u 是 odd；

已知 $t+u = $ odd，说明 s 是 even，因此 r 是 odd。

综上 s 和 t 为偶数。

答案 A

2. If $x = 2y+1$, and $y = 2w$, where w, x, and y are integers, which of the following must be an odd integer?

A. $xy + w$
B. $xy + w + 1$
C. $(x+y)w$
D. $wy + x$
E. $wx + y$

题干翻译

如果 $x = 2y+1$, $y = 2w$，其中 w, x, y 是整数，下列哪一项一定是奇数？

解题思路

根据 $x = 2y+1$ 可知 x 为 odd，

根据 $y = 2w$ 可知 y 为 even，

选项 D：$wy + x = $ even $+$ odd $= $ odd。

答案 D

3. If k and n are positive integers such that $k+n$ is an odd integer, which of the following must be an odd integer?

A. kn
B. $kn + n$
C. $kn + n^2$
D. $(k-1)^2 + n^2$
E. $k^2 + n$

题干翻译

如果 k 和 n 是正整数，$k+n$ 是奇数，那么下列哪一项一定是奇数？

解题思路

根据指数不改变奇偶性这一性质可知，当 $k+n=\text{odd}$ 时，k^2+n 一定为奇数。

答案 E

4. If x and y are both odd integers, which of the following must be a multiple of 4?
A. $x^2 y$ B. $x^2(y+1)$ C. $x(y+1)$ D. $(x-1)y^2$ E. $(x+1)(y-1)$

题干翻译

如果 x 和 y 都是奇数，下列哪一项一定是 4 的倍数？

解题思路

根据题干条件 x 和 y 为奇数，且相邻数字奇偶性相反，因此 $x+1$ 和 $y-1$ 为偶数，因此 E 选项 $(x+1)(y-1)=$ 偶数×偶数，为 4 的倍数。

答案 E

5. If both the sum and the product of the four positive integers a, b, c, and d are even, what is the greatest number of these integers that could be odd?
A. None B. One C. Two D. Three E. Four

题干翻译

如果 4 个正整数 a，b，c 和 d 的和与积都是偶数，那么这些整数中奇数个数最多为多少个？

解题思路

根据题干条件 $a \times b \times c \times d = \text{even}$ 可知，这 4 个数中至少有 1 个偶数，因此奇数个数小于等于 3 个。
根据题干条件 $a+b+c+d=\text{even}$ 可知，奇数的个数只能是偶数个。
综上，符合以上两个条件的奇数个数为小于等于 3 的偶数，这个范围下最大的数为 2。

答案 C

1.1.6 质数和合数

基本词汇

prime number 质数   composite number 合数   prime factor 质因数

概念

质数指的是有且仅有 1 和它本身两个正因数的正整数。20 以内，甚至 100 以内的质数大家都需要牢记。

合数指的是除了 1 和它本身以外，还存在其他正因数的正整数，即正因数至少含有 3 个正整数。

注意，质数和合数只涉及正整数，不包含负数和0；1既不是质数也不是合数。

在小学，我们学过因数分解这个概念，即把某一个正整数拆分成别的整数相乘的形式，比如：$12 = 1 \times 12 = 2 \times 6 = 3 \times 4$。所分解出来的1，2，3，4，6，12都是12这个数字的因数。在这些因数中，比如2，3，只能写成1×自身、2个正因数相乘的形式，即这些因数是质数，这类因数我们称之为质因数。

考点1　分解质因数

概念

把一个合数分解成若干个质数乘积的形式，叫作分解质因数。

例如：$12 = 2^2 \times 3$。

注意

为了方便做题，建议分解质因数的时候把结果写成如下形式：
(1) 把所有质因数从小到大排序。
(2) 相同的质因数合并成乘方的形式。

考点2　质数的性质

质数：只有1和自己本身两个正因数。

合数：除了1和自己本身之外，还有别的正因数，即至少含有3个因数。

注意，20以内，甚至100以内的质数我们必须牢记！考试经常会用到。

20以内的质数：2，3，5，7，11，13，17，19

20~100之间的质数：23，29，31，37，41，43，47，53，59，61，67，71，73，79，83，89，97。

在GRE数学考试中，质数还常考查两个性质：

（1）一个正整数如果刚好有且仅有3个正因数，那么它肯定是合数，同时它也一定是一个质数的平方。

例如：4有3个正因数：1，2，4，同时4是质数2的平方。

（2）**2是最小的质数，也是质数中唯一的偶数，其他的质数都是奇数。**

在GRE数学中，如果题目提到了一个质数不是2，也就是暗示这个数为奇数。

考点3　质因数和正因数的个数

请大家尝试求解一下36有多少个质因数和正因数。

首先，对36进行分解质因数：$36 = 2^2 \times 3^2$。

36分解出来了两个2和两个3，但是数学中规定质因数、因数重复只能算1次，所以36就只有两个质因数：2和3。

> **Tips**
> 质因数个数：分解质因数之后看底数，底数的个数＝质因数的个数。
> 正因数的个数：分解质因数之后看各项质因数的指数，各指数加1再相乘，得到的乘积就是正因数的个数。

如：$36 = 2^2 \times 3^2$，

质因数2的指数是2，质因数3的指数是2，所以36的正因数的个数为：$(2+1)(2+1) = 9$ 个。

If x is the product of the positive integers from 1 to 8, inclusive, and if i, k, m, and p are positive integers such that $x = 2^i 3^k 5^m 7^p$, then $i + k + m + p =$

A. 4　　　　　　B. 7　　　　　　C. 8　　　　　　D. 11　　　　　　E. 12

题干翻译

如果 x 是从1到8的正整数的乘积，并且如果 i，k，m 和 p 是正整数，$x = 2^i 3^k 5^m 7^p$，则 $i + k + m + p =$

解题思路

根据题干条件可知，$x = 1 \times 2 \times 3 \times 4 \times 5 \times 6 \times 7 \times 8$
$= 2 \times 3 \times 2^2 \times 5 \times 2 \times 3 \times 7 \times 2^3$
$= 2^7 \times 3^2 \times 5^1 \times 7^1 = 2^i 3^k 5^m 7^p$。

即 $i = 7$，$k = 2$，$m = 1$，$p = 1$，

因此 $i + k + m + p = 7 + 2 + 1 + 1 = 11$。

注意：如果题目中出现一个比较大的数值，通常需要我们对这个比较大的数值进行分解质因数。

答案 D

List P consists of the integers from 80 to 100, inclusive

Quantity A	Quantity B
The fraction of integers in list P that are prime	$\dfrac{1}{7}$

A. Quantity A is greater.
B. Quantity B is greater.
C. The two quantities are equal.
D. The relationship cannot be determined from the information given.

题干翻译

数组 P 由80到100的整数组成（包含首尾）。

解题思路

本题要求我们比较数组 P 中质数对应的比例与 $\frac{1}{7}$ 的大小关系。

首先我们要知道，连续数字 $x, \cdots, y$, inclusive, 元素的个数为 $y-x+1$。

因此，连续数字 80 到 100, inclusive 的情况下，元素个数 $=100-80+1=21$。

前面我们强调了，100 以内的质数一定要记牢。

80～100 中的质数为 83, 89, 97 这 3 个数字。

因此 80～100 中质数的比例 $=\frac{3}{21}=\frac{1}{7}$。

Tips

质数中只有 2 为偶数，其他质数都是奇数，因此只要考虑 80～100 里面的奇数就行。

个位以 5 结尾的为 5 的倍数，奇数中不用考虑以 5 结尾的数字。

易错数字：

87 为 3 的倍数，$87=3\times29$（可以利用 $8+7=15$，为 3 的倍数判断出来）。

91（$=13\times7$）这个数比较容易被误认为是质数，需特别记一下。

答案 C

What is the sum of the integers between 100 and 400 that have only 3 positive factors?

A. 579　　　　B. 650　　　　C. 771　　　　D. 819　　　　E. 940

题干翻译

100 到 400 之间只有 3 个正因数的整数之和是多少？

解题思路

一个数字如果只有 3 个正因数，那么这个数一定是质数的平方形式。

知道这个性质之后，我们就会知道本题主要是想让我们找出 10 到 20 之间的质数（正如我们之前强调的，大家一定要牢记 20 以内的质数），算出其平方对应的数值，再汇总求和即可。

10 到 20 之间的质数：11, 13, 17, 19,

因此，100 到 400 之间只有三个正因数的数字分别是：$11^2, 13^2, 17^2, 19^2$。

$11^2+13^2+17^2+19^2=940$。

答案 E

Quantity A	Quantity B
The product of the positive prime factors of 30	The sum of the positive prime factors of 30

A. Quantity A is greater.

B. Quantity B is greater.

C. The two quantities are equal.

D. The relationship cannot be determined from the information given.

解题思路

本题让我们比较的是 30 的质因数的乘积和 30 的质因数的和的大小关系。

先对 30 进行分解质因数：$30 = 2 \times 3 \times 5$，

由此可知，30 的质因数是 2，3，5。

数量 A：30 的质因数的乘积 $= 2 \times 3 \times 5 = 30$；

数量 B：30 的质因数的和 $= 2 + 3 + 5 = 10 (<30)$，因此，数量 A 更大。

答案　A

a and b are distinct odd prime numbers.

　　　　　Quantity A　　　　　　　　　　　　Quantity B

The number of positive factors of $2ab^2$　　The number of positive factors of a^2b^3

A. Quantity A is greater.

B. Quantity B is greater.

C. The two quantities are equal.

D. The relationship cannot be determined from the information given.

题干翻译

a 和 b 为不同的质数，且 ab 为奇数。

解题思路

本题要我们比较的是 $2ab^2$ 正因数的个数与 a^2b^3 正因数的个数的大小关系。

正因数的个数 = 分解质因数之后各项质因数的指数加 1 再相乘得到的乘积。

根据题干条件 a 和 b 为不同的质数且为奇数可知，$2ab^2$ 和 a^2b^3 就是分解质因数的最终形式。

因此数量 A：$2ab^2$，其正因数的个数 $= (1+1)(1+1)(2+1) = 12$；

因此数量 B：a^2b^3，其正因数的个数 $= (2+1)(3+1) = 12$。

数量 A = 数量 B。

答案　C

练 习

1. If q and r are prime numbers greater than 3 and $qr^2 < 450$, what is the greatest possible value of r?

   A. 5 B. 7 C. 11 D. 13 E. 17

2. What is the product of the different positive prime factors of 3,528?

   A. 6 B. 14 C. 42 D. 294 E. 441

3. p is a prime number greater than 3.

Quantity A	Quantity B
The number of positive divisors of $2p$	The number of positive divisors of p^2

   A. Quantity A is greater.
   B. Quantity B is greater.
   C. The two quantities are equal.
   D. The relationship cannot be determined from the information given.

4. If $n = ab$, where a and b are different prime numbers, which of the following statements must be true? Indicate all such statements.

   A. $\sqrt{n}$ is an integer.
   B. $n + 1$ is a prime number.
   C. n^2 has 9 different positive factors.

5. Positive integer n is the product of 4 prime numbers. When n is divided by 77, the result is a multiple of 5. Which of the following could be the quotient when n is divided by 7?

   A. 110 B. 220 C. 330 D. 440 E. 550

6. A positive integer with exactly two different divisors greater than 1 must be

   A. a prime number
   B. an even integer
   C. a multiple of 3
   D. the square of a prime number
   E. the square of an odd integer

答案及解析

1. If q and r are prime numbers greater than 3 and $qr^2 < 450$, what is the greatest possible value of r?

A. 5　　　　　B. 7　　　　　C. 11　　　　　D. 13　　　　　E. 17

题干翻译

如果 q 和 r 都是大于 3 的质数，且 $qr^2 < 450$，r 最大可能的值是多少？

解题思路

要求 r 的最大可能值，那么 q 就应该尽量小。

由于 q 和 r 都是大于 3 的质数，所以 q 最小取 5，

那么 r^2 小于 90，满足要求的 r 为 7。

答案　B

2. What is the product of the different positive prime factors of 3,528?

A. 6　　　　　B. 14　　　　　C. 42　　　　　D. 294　　　　　E. 441

题干翻译

3528 的不同的正质因数的乘积是多少？

解题思路

$3528 = 2^3 \times 3^2 \times 7^2$，

因此 3528 的质因数为 2，3，7。

3528 的正质因数的乘积 $= 2 \times 3 \times 7 = 42$。

答案　C

3. p is a prime number greater than 3.

Quantity A	Quantity B
The number of positive divisors of $2p$	The number of positive divisors of p^2

A. Quantity A is greater.
B. Quantity B is greater.
C. The two quantities are equal.
D. The relationship cannot be determined from the information given.

题干翻译

p 是大于 3 的质数。

解题思路

本题要我们比较的是 $2p$ 正因数的个数和 p^2 正因数的个数的大小。

正因数的个数 = 分解质因数之后各项质因数的指数加 1 再相乘得到的乘积。

根据题干条件 "p 为大于 3 的质数" 可知，$2p$ 和 p^2 为分解质因数的最终形式。

数量 A：$2p$ 正因数的个数 = $(1+1)(1+1) = 4$；

数量 B：p^2 正因数的个数 = $2+1 = 3 (<4)$。

因此数量 A 更大。

答案 A

4. If $n = ab$, where a and b are different prime numbers, which of the following statements must be true? Indicate <u>all</u> such statements.

A. $\sqrt{n}$ is an integer.

B. $n + 1$ is a prime number.

C. n^2 has 9 different positive factors.

题干翻译

如果 $n = ab$，其中 a 和 b 是不同的质数，下列哪个陈述一定为真？

解题思路

A 选项：$\sqrt{n}$ 是一个整数，根据题干条件 "a 与 b 为不同的质数" 可知，$n = ab$ 不是一个平方数，因此 $\sqrt{n}$ 不是整数，A 选项错误。

B 选项：$n+1$ 是一个质数，如果 $a = 3$，$b = 5$，则 $n+1 = 16$ 为合数，不是质数，B 选项错误。

C 选项：n^2 有 9 个正因数，正因数的个数 = 分解质因数之后各项质因数的指数加 1 再相乘得到的乘积，n^2 分解质因数的最终形式为 a^2b^2，因此 n^2 的正因数个数 = $(2+1)(2+1) = 9$，C 选项正确。

答案 C

5. Positive integer n is the product of 4 prime numbers. When n is divided by 77, the result is a multiple of 5. Which of the following could be the quotient when n is divided by 7?

A. 110　　　　B. 220　　　　C. 330　　　　D. 440　　　　E. 550

题干翻译

正整数 n 是 4 个质数的乘积。当 n 除以 77 时，结果是 5 的倍数。当 n 除以 7 时，下列哪一项可能是它的商？

解题思路

由题干可知，n 的质因数应包含 7，11，5，以及另外一个其他的质数。

那么 n 除以 7 之后，得到的结果应包含 11，5，以及另外一个其他的质数。
即 $110 = 11 \times 5 \times 2$，符合要求。

答案　A

6. A positive integer with exactly two different divisors greater than 1 must be
　A. a prime number　　　　　　　B. an even integer
　C. a multiple of 3　　　　　　　D. the square of a prime number
　E. the square of an odd integer

题干翻译
一个正整数有且仅有两个比 1 大的因数，问这个正整数一定是什么？

解题思路
注意，1 是任何数字的因数。
题干的条件相当于告诉我们，这个正整数除了 1 以外，还有 2 个正因数，即这个正整数有 3 个正因数。
根据性质：一个正整数有且仅有 3 个正因数，它一定是质数的平方。
因此 D 选项正确。

答案　D

1.1.7　连续整数

基本词汇

consecutive integers　连续整数

概念

连续整数是指公差等于 1 的递增的等差数列，比如 1，2，3，4，5 属于连续整数，但是 5，4，3，2，1 不属于连续整数。
连续奇整数指的是公差等于 2 的递增的等差数列，比如 1，3，5，7 属于连续奇整数。
连续偶整数指的是公差等于 2 的递增的等差数列，比如 2，4，6，8 属于连续偶整数。

考点

连续整数的性质
在 GRE 数学考试中，连续整数的考点非常单一，主要考查以下几个性质。
（1）2 个连续整数的乘积一定是偶数（2 的倍数）。
（2）3 个连续整数的乘积一定是 6 的倍数。
（3）2 个连续偶数的乘积一定是 8 的倍数。

例题 01.

j, k, and e are three positive consecutive integers, and $j < k < e$.

Quantity A	Quantity B
jke	k^3

A. Quantity A is greater.

B. Quantity B is greater.

C. The two quantities are equal.

D. The relationship cannot be determined from the information given.

题干翻译

j, k, e 为连续正整数，且 $j < k < e$。

解题思路

连续整数是指公差等于 1 的递增的等差数列。

根据题干条件"j, k, e 为连续正整数，且 $j < k < e$"可知，

$j = k - 1$, $e = k + 1$，

数量 A $= jke = (k-1)(k)(k+1) = k^3 - k$。

因为 $k > 0$，

因此 $k^3 - k < k^3$（$k^3 =$ 数量 B），

即数量 B 更大。

答案 B

例题 02.

If n is an integer, then $n^3 - n$ must be divisible by which of the following?

A. 4   B. 5   C. 6   D. 7   E. 8

题干翻译

如果 n 是整数，那么 $n^3 - n$ 一定可以被以下哪一项整除？

解题思路

$n^3 - n = n(n^2 - 1) = n(n-1)(n+1)$。

$n(n-1)(n+1)$ 是 3 个连续整数相乘，乘积肯定是 2 的倍数、3 的倍数和 6 的倍数。

即 $n^3 - n$ 肯定能够被 2, 3, 6 整除。

答案 C

k is an integer.

Quantity A	Quantity B
The remainder when $k^2 - k$ is divided by 2	0

A. Quantity A is greater.

B. Quantity B is greater.

C. The two quantities are equal.

D. The relationship cannot be determined from the information given.

题干翻译

k 是一个整数。

解题思路

本题让我们比较的是 $k^2 - k$ 除以 2 的余数与 0 的大小关系。

$k^2 - k = k(k-1)$，

$k(k-1)$ 为两个连续整数相乘，乘积肯定是 2 的倍数，因此 $k^2 - k$ 可以被 2 整除。

即 $k^2 - k$ 除以 2 余 0。

答案 C

练 习

1. x, y, and z are consecutive positive integers and $x < y < z$.

Quantity A	Quantity B
$\dfrac{xy}{z}$	$z - \dfrac{1}{z}$

A. Quantity A is greater. B. Quantity B is greater.

C. The two quantities are equal. D. The relationship cannot be determined from the information given.

2. q, r, and s are consecutive positive integers and $q < r < s$.

Quantity A	Quantity B
$\dfrac{qs}{r}$	$r - \dfrac{1}{r}$

A. Quantity A is greater.

B. Quantity B is greater.

C. The two quantities are equal.

D. The relationship cannot be determined from the information given.

3. n is a positive integer and $p = n(n+1)(n+2)$.

Quantity A | Quantity B
The remainder when p is divided by 3 | 1

A. Quantity A is greater.

B. Quantity B is greater.

C. The two quantities are equal.

D. The relationship cannot be determined from the information given.

答案及解析

1. x, y, and z are consecutive positive integers and $x < y < z$.

Quantity A | Quantity B
$\dfrac{xy}{z}$ | $z - \dfrac{1}{z}$

A. Quantity A is greater.

B. Quantity B is greater.

C. The two quantities are equal.

D. The relationship cannot be determined from the information given.

题干翻译

x, y, z 为连续正整数，且 x < y < z。

解题思路

本题让我们比较的是 $\dfrac{xy}{z}$ 与 $z - \dfrac{1}{z}$ 的大小关系。

连续整数是指公差等于1的递增的等差数列。

根据题干条件"x, y, z 为连续正整数，且 $x<y<z$"可知，
$x = z-2$，$y = z-1$，
数量 A $= \dfrac{xy}{z} = \dfrac{(z-2)(z-1)}{z} = \dfrac{z^2-3z+2}{z}$。

数量 A − 数量 B $= \dfrac{z^2-3z+2}{z} - \left(z - \dfrac{1}{z}\right)$,

$\qquad\qquad\qquad\quad = \dfrac{z^2-3z+2}{z} - \dfrac{z^2-z}{z}$,

$\qquad\qquad\qquad\quad = \dfrac{-2z+2}{z}$。

根据题干条件"x 为正整数"可知，$x=z-2>0$，$z>2$，

因此，数量 A − 数量 B $= \dfrac{-2z+2}{z} < 0$。

故数量 B 更大。

答案　B

2. q, r, and s are consecutive positive integers and $q<r<s$.

Quantity A	Quantity B
$\dfrac{qs}{r}$	$r - \dfrac{1}{r}$

A. Quantity A is greater.

B. Quantity B is greater.

C. The two quantities are equal.

D. The relationship cannot be determined from the information given.

题干翻译

q, r, s 为连续正整数，且 $q<r<s$。

解题思路

本题让我们比较的是 $\dfrac{qs}{r}$ 与 $r-\dfrac{1}{r}$ 的大小关系。

连续整数是指公差等于 1 的递增的等差数列。

根据题干条件"q, r, s 为连续正整数，且 $q<r<s$"可知，$q=r-1$，$s=r+1$。

数量 A $= \dfrac{qs}{r} = \dfrac{(r-1)(r+1)}{r} = \dfrac{r^2-1}{r} = r - \dfrac{1}{r} =$ 数量 B。

答案　C

3. n is a positive integer and $p = n(n+1)(n+2)$.

Quantity A	Quantity B
The remainder when p is divided by 3	1

A. Quantity A is greater.

B. Quantity B is greater.

C. The two quantities are equal.

D. The relationship cannot be determined from the information given.

题干翻译

n 为正整数，且 $p = n(n+1)(n+2)$。

解题思路

根据题干条件 $p = n(n+1)(n+2)$ 可知，p 为3个连续整数的乘积，p 为6的倍数。

由此可知，p 可以被3整除，即 p 除以3的余数 $0<1$，

因此，数量B更大。

答案　B

1.1.8 数个数

在GRE考试中，有时候会涉及项数的计算。

我们一起来思考一下，连续正整数1到100（包含首尾），一共有多少个数字呢？

答案是100个。

> **Tips**
>
> 涉及连续数字（公差等于1的等差数列）个数计算问题，大家应该知道以下几个知识点。
>
> 连续数字 $x, x+1, \cdots, y$, inclusive 的项数为 $y-x+1$（inclusive 指的是包含首尾项）。
>
> 连续数字 $x, x+1, \cdots, y$, exclusive 的项数为 $y-x+1-2$ 即 $y-x-1$（exclusive 指的是不包含首尾项）。

如果不是连续数字，而是等差数列，项数应该如何计算呢？

比如，连续奇数 $1, 3, 5, \cdots, 99$, inclusive 有多少项呢？

答案是50项。

公差为 d 的等差数列 $x, x+d, \cdots, y$ 的项数有多少呢？

$x, x+d, \cdots, y$, inclusive 的项数为 $\frac{y-x}{d}+1$。

$x, x+d, \cdots, y$, exclusive 的项数为 $\frac{y-x}{d}+1-2 = \frac{y-x}{d}-1$。

如果不是等差数列，项数又应该如何计算呢？

比如：-100 到 100 中有多少个3的倍数？

这种情况下，我们应该先找出在这个范围下等差数列的最小项和最大项。

项数 $= \frac{\text{等差数列的最大项} - \text{等差数列的最小项}}{\text{公差}} + 1$。

-100 到 100 中，3的倍数的最小项为 -99，最大项为 99，

因此，-100 到 100 中3的倍数个数 $= \frac{99-(-99)}{3}+1 = 67$。

How many integers are multiples of 7 from −42 to 98, inclusive?

题干翻译

−42 到 98 中（包含首尾），7 的倍数有多少个？

解题思路

对于等差数列而言，项数 = $\dfrac{\text{等差数列的最大项} - \text{等差数列的最小项}}{\text{公差}} + 1$，

−42 和 98 都为 7 的倍数，

因此，7 的倍数的个数 = $\dfrac{98 - (-42)}{7} + 1 = 21$。

答案　21

How many integers are multiples of 3 from 100 to 10,000, inclusive?

题干翻译

100 到 10000 中（包含首尾），3 的倍数有多少个？

解题思路

由于 100 和 10000 都不是 3 的倍数，因此我们要先找出这个范围内 3 的倍数的最小值和最大值：
3 的倍数的最小值为 102，3 的倍数的最大值为 9999。

因此，100 到 10000 中 3 的倍数的个数 = $\dfrac{9999 - 102}{3} + 1 = 3300$。

答案　3300

1.2 分数和小数

1.2.1 分数

基本词汇

fraction 分数
numerator 分子
denominator 分母

rational number 有理数
nonzero integer 非零整数

reciprocal 倒数
mixed number 带分数

概念

分数的形式为 $\dfrac{a}{b}$，其中 a 和 b 都是整数，且 $b \neq 0$。

其中整数 a 叫作分子，b 叫作分母。例如，$\dfrac{1}{3}$ 是一个分数，1 是分子，3 是分母。这类分数，即能够写成两个整数相除形式的数，也被称为有理数。

如果分子 a 与分母 b 同时乘以一个相同的非零整数，所形成的分数与 $\dfrac{a}{b}$ 相等。

例如：$\dfrac{1}{3} = \dfrac{(1)(4)}{(3)(4)} = \dfrac{4}{12}$。

如果分数的分子或分母上有负号，这个负号可以移动到分数的前面。

例如：$\dfrac{-1}{3} = \dfrac{1}{-3} = -\dfrac{1}{3}$。

如果分数的分子和分母有共同的因数，那么分子和分母可以分解因数，将共同的因数约分，形成的结果与原来的分数相等。

例如：$\dfrac{4}{12} = \dfrac{(1)(4)}{(3)(4)} = \dfrac{1}{3}$。

考点 分数运算

1. 加减运算

当两个分数进行加减运算时，如果两个分数的分母相同，则分母保持不变，直接将分子相加减。

例如：$\dfrac{1}{5} + \dfrac{2}{5} = \dfrac{1+2}{5} = \dfrac{3}{5}$。

当两个分数进行加减运算时，如果两个分数的分母不相同，那么要先找出两个分数分母的公倍数（一般找最小公倍数），然后将两个分数转化为以最小公倍数为分母的分数，再将分子相加减。

例如：$\frac{1}{3}+\frac{2}{5}=\frac{(1)(5)}{(3)(5)}+\frac{(2)(3)}{(5)(3)}=\frac{5}{15}+\frac{6}{15}=\frac{11}{15}$。

2. 乘法运算

当两个分数相乘，直接将两个分数的分子部分相乘，形成新的分子，将两个分数的分母部分相乘，形成新的分母。

例如：$\frac{1}{3}\times\frac{2}{5}=\frac{(1)(2)}{(3)(5)}=\frac{2}{15}$。

3. 除法运算

当一个分数除以另一个分数，首先将第二个分数写成其对应的倒数形式，然后再用第一个分数乘以第二个分数的倒数。

例如：$\frac{17}{8}\div\frac{3}{5}=\left(\frac{17}{8}\right)\times\left(\frac{5}{3}\right)=\frac{85}{24}$。

如果一个分数，例如 $4\frac{1}{8}$，既有整数部分又有分数部分，那么这个分数被称为带分数。带分数 $4\frac{1}{8}=4+\frac{1}{8}$。将一个带分数变成普通的分数形式，首先将整数部分转换成与之等价的分数形式，然后再与分数部分相加。

例如：$4\frac{1}{8}=4+\frac{1}{8}=\left(\frac{4}{1}\right)\times\left(\frac{8}{8}\right)+\frac{1}{8}=\frac{32}{8}+\frac{1}{8}=\frac{33}{8}$。

例题

$\frac{a}{b}=1$

Quantity A   Quantity B

$\frac{a+b}{a}$ $a+b$

A. Quantity A is greater.
B. Quantity B is greater.
C. The two quantities are equal.
D. The relationship cannot be determined from the information given.

解题思路

本题让我们比较 $\frac{a+b}{a}$ 与 $a+b$ 的大小关系。

根据题干条件 $\frac{a}{b}=1$,可知 $a=b$,

因此,数量 A $= \frac{a+b}{a} = \frac{2a}{a} = 2$,

数量 B $= a+b = 2a$,

由于题干没有提供 a 与 1 的大小关系的信息,因此数量 A 与数量 B 的数值大小关系我们无法确定。

答案 D

1.2.2 小数

基本词汇

decimal 小数　　　　　　　digit 数位上的数值　　　　　scientific decimal notation
decimal notation 十进制　　terminating decimal 有限小数　科学计数法
decimal point 小数点　　　 irrational number 无理数　　　repeating decimal 循环小数

考点 1　小数的数位

GRE 考试中的小数体系是基于十进制的。首先,我们来认识一下各个数位的表达。

例如:1234.567

1:thousands 千位　　　　　　　　　　　".":decimal point 小数点
2:hundreds 百位　　　　　　　　　　　5:tenths 十分位
3:tens 十位　　　　　　　　　　　　　 6:hundredths 百分位
4:units or ones 个位　　　　　　　　　 7:thousandths 千分位

大家可以发现,整数部分的单词都是由实数词后面直接加 s 构成,而小数部分的单词,都是序数词后面直接加 s。

注意,整数部分是从"个位"开始的,但小数部分是从"十分位"开始的。

在 GRE 数学考试中,数位问题主要考查读题能力,需要考生了解数位对应的具体数值和位置:

一个数字在个位上,代表它自己;
一个数字在十位上,代表这个数字 ×10;
一个数字在百位上,代表这个数字 ×100;
一个数字在十分位上,代表这个数字 ×0.1;
一个数字在百分位上,代表这个数字 ×0.01;
……

例如:123.45 代表的实际数值是 $1 \times 100 + 2 \times 10 + 3 \times 1 + 4 \times 0.1 + 5 \times 0.01$。

考点2　小数的四舍五入

1. 表达形式

常见的表达形式：

round to 四舍五入，

to the nearest 四舍五入，

round up to 只入不舍，

round down to 只舍不入。

例如：round to hundreds 四舍五入到百位（即将百位后面的数位——十位和个位化为0）。

四舍五入到某一数位，我们考虑后面紧靠的数位的情况：

如果后面紧靠的数位上的数值≥5，则后面的数值向前进1；

如果后面紧靠的数位上的数值<5，则后面的数值直接舍去。

例如：

1774，round to hundreds。四舍五入到百位，则考虑百位后面十位的情况，因为十位的数值 $7>5$，所以向前进1，得到结果1800。

1774，round to tens。四舍五入到十位，则考虑十位后面个位的情况，因为个位的数值 $4<5$，所以直接舍掉个位，得到结果1770。

round up to 指的是"只入不舍"，只入不舍到某一数位，考虑后面一位数的情况，但是后面的数值不管是什么，直接向前进1。

round down to 指的是"只舍不入"，只舍不入到某一数位，考虑后面一位数的情况，但是后面的数值不管是什么，直接舍去。

例如：

9.117，round up to tenths 的结果是9.2；

9.117，round down to hundredths 的结果是9.11。

注意，当涉及实际场景话题，让我们计算人数或商品数量的时候，可能会隐蔽地考查只入不舍或只舍不入这一知识点，因为人数和商品数只能取整数。

例如，为了保证某项工作在两天内必须完成，我们计算出需要9.1个人，那么在实际雇人的时候我们必须雇10个人。

2. 考法

四舍五入的考法，可能会考查一个数值四舍五入后的结果；也可能会告诉我们四舍五入的结果，让我们反推这个数值对应的取值范围。

考点3　小数的科学计数法

科学计数法即用 $a \times 10^n$ 这种形式（a 是一个 $1\sim10$ 之间的数字）来表示一个比较复杂的数值。

比如：

$1234000 = 1.234 \times 10^6$；$0.001234 = 1.234 \times 10^{-3}$。

在 GRE 考试中，我们可能会涉及科学计数法表达式相关的计算，或者用科学计数法表达比较复杂的数字。这个考点难度不大。

考点 4　有限小数

如果小数点右边有有限个数字，则称为有限小数。

我们可以很容易将有限小数转换成分子和分母都是整数的等价分数。例如，$1.3 = 1 + \frac{3}{10} = \frac{13}{10}$。

有限小数的考法通常是题目给出一个分数形式 $\frac{b}{a}$，让我们来判断这个分数是否有限。

如果分子 ÷ 分母，得到的结果是有限长度，则为有限小数；如果分子 ÷ 分母，得到的结果是无限长度，则为无限小数。

GRE 考试中，虽然系统自带计算器，但系统自带的计算器只能显示 8 个数字，因此在判断分数是否是有限小数时，题目给出的式子最后转换成的小数数位往往会超过 GRE 系统自带的计算器能够显示的范围，因此这类题目我们需要利用技巧来判断，而不是使用计算器。

那么我们究竟如何来判断 $\frac{b}{a}$ 是不是有限小数呢？

先保证将分数化为最简形式，即分子和分母除了 1 以外不存在其他的公因数，然后对分母进行分解质因数，只要分解的结果只含有 2 或 5，那么就是有限小数；如果分解的结果含有 2 和 5 之外的质因数，则为无限小数。

考点 5　循环小数

当一个分数的分子和分母都是整数的时候，这个分数写成小数的时候可能是有限小数，也可能是循环小数。例如，$\frac{1}{2} = 0.5$，是一个有限小数；$\frac{1}{3} = 0.333\cdots = 0.\overline{3}$，是一个循环小数。在 GRE 数学考试中，循环小数一般是在所循环的数字上加一个横线，这个横线称为循环节。

在 GRE 中，我们可能会遇到要求我们将循环小数转换成分数形式的题目，因此我们需要了解一下循环小数转换成分数的规则：

①分数分母部分由数字 9 构成，9 的个数 = 循环节下数字的个数。

②分数分子部分的数值 = 循环节下数字的数值。

比如，将循环小数 $0.\overline{3}$ 转换成分数：

分数分母部分由数字 9 构成，9 的个数 = 循环节下数字的个数：循环小数 $0.\overline{3}$ 的循环节下只有 1 个数字 3，因此这个小数对应的分数分母部分为 1 个 9。

分数分子部分的数值 = 循环节下数字的数值：循环小数 $0.\overline{3}$ 的循环节下的数字为 3，因此分子部分为 3。

综上，$0.\overline{3}$ 对应的分数形式为 $\frac{3}{9}$，得到这个结果之后我们也可以进一步化简，得到 $\frac{1}{3}$。

考点6 无限不循环小数

不是所有的小数都是有限小数或循环小数，例如，$\sqrt{2}=1.41421356237\cdots$，这就是一个无限不循环小数。这种小数点后的数字有无限多个，且不会循环的小数叫作无理数。常见的无理数包含非完全平方数、π 等。

Quantity A	Quantity B
The digit in the hundredths place in the number 0.083	The digit in the hundreds place in the number 1,600

A. Quantity A is greater.

B. Quantity B is greater.

C. The two quantities are equal.

D. The relationship cannot be determined from the information given.

解题思路

本题让我们比较的是 0.083 百分位对应的数值与 1600 百位对应的数值。

0.083 百分位对应的数值为 8，1600 百位对应的数值为 6，

$8>6$，

故数量 A 更大。

答案 A

If the tens digit x and units digit y of a positive integer n are reversed, the resulting integer is 9 more than n. What is y in terms of x?

A. $10-x$ B. $9-x$ C. $x+9$ D. $x-1$ E. $x+1$

题干翻译

正整数 n 的十位数 x 和个位数 y 进行互换，互换之后得到的数值比 n 大 9。问如何用 x 表示 y？

解题思路

根据题干可知 $n=xy=10x+y$（x，y 代表数位上的数值），互换之后的数值为 $yx=10y+x$。

根据题干条件，互换后的结果比 n 大 9，我们可以得到：

$10x+y+9=10y+x$，

$9y=9x+9$，

$y=x+1$。

答案 E

例题 03.

The sides of a square region, measured to the nearest centimeters, are 9 centimeters long. The least possible value of the actual area of the square region is

A. 81.00 sqcm B. 78.75 sqcm
C. 75.75 sqcm D. 72.25 sqcm
E. 56.25 sqcm

题干翻译

正方形区域的边长，四舍五入到整厘米为9厘米。这个正方形区域实际面积的最小可能值为多少？

解题思路

边长四舍五入到整厘米的结果为9，说明边长 x 的取值范围应该是 $8.5 \leqslant x < 9.5$。

正方形的面积 = 边长^2，

因此，正方形面积最小，意味着边长最小。

边长最小为8.5，因此正方形面积值的最小可能 = $8.5^2 = 72.25$ sqcm。

答案 D

例题 04.

To determine her state income tax last year, Elena computed 5 percent of her gross income, rounded the resulting figure down to the nearest $100, and subtracted $50 for each dependent that she claimed. If Elena's gross income was $35,500 and she claimed 2 dependents, how much did she determine her state income tax to be?

A. $1,600 B. $1,625 C. $1,650 D. $1,675 E. $1,700

题干翻译

为了确定她去年的收入所得税，Elena 计算了她总收入的5%，将结果四舍不入到整百，并且对于她申报的每个受抚养人减去50美元。如果 Elena 的总收入是35500美元，她申报了2个受抚养人，那么她的收入所得税是多少？

解题思路

首先用 Elena 的收入 $\times 5\% = 35500 \times 5\% = 1775$。

再根据题干要求对这个数字进行只舍不入到整百，$1775 \approx 1700$。

$1700 - 50 \times 2 = 1600$。

答案 A

例题 05.

A gardener plans to cover a rectangular plot of land with pine bark mulch to a depth of 4 inches. The plot measures 8 feet by 12 feet, and the gardener will buy mulch packed in bags. If each bag contains 3.5 cubic feet of mulch and costs $6, what is the cost of the least number of bags that the gardener will need to cover the plot? (Note: 1 foot = 12 inches.)

A. $30　　　B. $48　　　C. $54　　　D. $55　　　E. $60

题干翻译

一名园丁计划用松树皮覆盖物覆盖一块长方形的土地，深度为4英寸。这块地长8英尺，宽12英尺，园丁会以袋装的形式购买覆盖物。如果每袋包含3.5立方英尺的覆盖物，每袋价格为6美元，那么园丁覆盖这块地所需的最少袋覆盖物的成本是多少？（注：1英尺=12英寸。）

解题思路

根据题干条件可知，这片区域的体积 $=\dfrac{4}{12}\times 8\times 12=32\ \text{feet}^3$。

需要购买的材料袋数 $=\dfrac{32}{3.5}\approx 10$ 袋（请注意，商品数量只能取整，而且为了保证能把区域完全覆盖，我们必须进行只入不舍的操作），

因此，园丁覆盖这片区域的最低成本 $=10\times 6=60$ 美元。

答案 E

例题 06.

$(2.82\times 10^{-51})-(3.96\times 10^{-49})=$

A. -3.9318×10^{-49}　　　B. -1.7804×10^{-51}　　　C. -1.14×10^{-100}

D. 1.7804×10^{-51}　　　E. 3.9318×10^{-49}

解题思路

这题其实可以不需要真的去计算出结果，它想考查的是学生如何利用性质，用最短的时间把正确选项选出来。

方法1：

首先观察选项，我们会发现选项都是几点几（10以内的数字）乘以 10^{-n} 这种形式。

提取公因数，应该提取的是 10^{-49}，扫读选项发现以 10^{-49} 为公因数的选项只有A选项和E选项，而这两个选项的区别在于正负。因此，我们只需要判断题干表达式的正负就行。

显然 $2.82\times 10^{-51}<3.96\times 10^{-49}$，

所以 $(2.82\times 10^{-51})-(3.96\times 10^{-49})$ 的结果为负数，

故选A。

方法 2:

$(2.82 \times 10^{-51}) - (3.96 \times 10^{-49})$
$= 10^{-49}(2.82 \times 10^{-2} - 3.96)$
$= 10^{-49}(0.0282 - 3.96)$
$= -3.9318 \times 10^{-49}$

答案 A

Which of the following fractions has a decimal equivalent that is a terminating decimal?

A. $\dfrac{11}{189}$ B. $\dfrac{17}{196}$ C. $\dfrac{19}{225}$ D. $\dfrac{35}{144}$ E. $\dfrac{39}{128}$

题干翻译

以下哪个分数写成小数形式，结果是有限小数？

解题思路

由于本题中 5 个选项的分数形式均为最简形式，因此我们直接对选项的分母进行分解质因数即可。

$189 = 3^3 \times 7$，分解的结果有 2，5 之外的质因数 3 和 7，不是有限小数；

$196 = 2^2 \times 7^2$，分解的结果有 2，5 之外的质因数 7，不是有限小数；

$225 = 3^2 \times 5^2$，分解的结果有 2，5 之外的质因数 3，不是有限小数；

$144 = 2^4 \times 3^2$，分解的结果有 2，5 之外的质因数 3，不是有限小数；

$128 = 2^7$，分解的质因数只有 2，是有限小数。

答案 E

The repeating decimal $1.\overline{ab}$, where a and b are different digits, is equivalent to the fraction $\dfrac{n}{d}$, where n and d are positive integers whose greatest common factor is 1. What is the greatest possible value of $n + d$?

A. 296   B. 297   C. 298   D. 299   E. 301

题干翻译

循环小数 $1.\overline{ab} = \dfrac{n}{d}$，其中 a 和 b 是不同的数字，n 和 d 为正整数，且最大公因数为 1。$n + d$ 的最大可能值是多少？

解题思路

$1.\overline{ab} = 1 + 0.\overline{ab}$

当循环小数转换成分数时，

①分数分母部分由数字 9 构成，9 的个数 = 循环节下数字的个数，

②分数分子部分的数值 = 循环节下数字的数值。

对于循环小数 $0.\overline{ab}$ 来说，它对应的分母部分为 99，对应的分子部分为 ab，

即 $0.\overline{ab} = \dfrac{ab}{99}$，

那么 $1.\overline{ab} = 1 + 0.\overline{ab} = 1 + \dfrac{ab}{99} = \dfrac{99 + ab}{99} = \dfrac{n}{d}$。

由于题干要我们算 $n + d$ 的最大可能值，

那么 ab 要最大，且 $99 + ab$ 要与 99 互质，因此 d 最大为 99，n 最大为 $99 + 98 = 197$。

综上，$n + d$ 的最大可能值 $= 99 + 197 = 296$。

答案 A

例题 09.

For which of the following values of x does the expression $\sqrt{\dfrac{24}{x+1}}$ represents an irrational number?

A. 12.5　　　　B. 36.5　　　　C. 53　　　　D. 71　　　　E. 95

题干翻译

x 取哪个值，表达式 $\sqrt{\dfrac{24}{x+1}}$ 是一个无理数？

解题思路

无理数，也称为无限不循环小数，不能写作两整数之比。若将其写成小数形式，小数点之后的数字有无限多个，并且不会循环。

A：$x = 12.5$　$\sqrt{\dfrac{24}{x+1}} = \sqrt{\dfrac{24}{13.5}} = \sqrt{\dfrac{48}{27}} = \sqrt{\dfrac{16}{9}} = \dfrac{4}{3}$（为整数比，不符合无理数定义）。

B：$x = 36.5$　$\sqrt{\dfrac{24}{x+1}} = \sqrt{\dfrac{24}{37.5}} = \sqrt{\dfrac{48}{75}} = \sqrt{\dfrac{16}{25}} = \dfrac{4}{5}$（为整数比，不符合无理数定义）。

C：$x = 53$　$\sqrt{\dfrac{24}{x+1}} = \sqrt{\dfrac{24}{54}} = \sqrt{\dfrac{4}{9}} = \dfrac{2}{3}$（为整数比，不符合无理数定义）。

D：$x = 71$　$\sqrt{\dfrac{24}{x+1}} = \sqrt{\dfrac{24}{72}} = \sqrt{\dfrac{1}{3}} = \dfrac{1}{\sqrt{3}}$（不是整数比，符合无理数定义）。

E：$x = 95$　$\sqrt{\dfrac{24}{x+1}} = \sqrt{\dfrac{24}{96}} = \sqrt{\dfrac{1}{4}} = \dfrac{1}{2}$（为整数比，不符合无理数定义）。

答案 D

练 习

1. $w = 26.956$

Quantity A　　　　　　　　　　　　　　Quantity B

The digit in the hundredths place of w　　The digit in the hundreds place of $4w$

A. Quantity A is greater.
B. Quantity B is greater.
C. The two quantities are equal.
D. The relationship cannot be determined from the information given.

2. The two-digit positive integer m is equal to twice the sum of its digit.

Quantity A　　　Quantity B
　m　　　　　　20

A. Quantity A is greater.
B. Quantity B is greater.
C. The two quantities are equal.
D. The relationship cannot be determined from the information given.

3. The speed of light is 3×10^8 meters per second, rounded to the nearest 10^8 meters per second. A "light-hour" is the distance that light travels in an hour.

Quantity A　　　　　　　　　　　　　　Quantity B

The number of kilometers in a light-hour　　10^{10}

A. Quantity A is greater.
B. Quantity B is greater.
C. The two quantities are equal.
D. The relationship cannot be determined from the information given.

4. A certain brand of chewing gum can be purchased in only two ways:

1 piece at a time at 7 cents a piece

5 pieces at a time at 30 cents for the 5 pieces

Quantity A　　　　　　　　　　　　　　Quantity B

The greatest number of pieces of the gum that can be purchased with 87 cents　　The greatest number of pieces of the gum that can be purchased with 82 cents

A. Quantity A is greater.

B. Quantity B is greater.

C. The two quantities are equal.

D. The relationship cannot be determined from the information given.

5. $10^2(0.00839) =$

A. 8.39×10^{-4}

B. 8.39×10^{-3}

C. 8.39×10^{-2}

D. 8.39×10^{-1}

E. 8.39×10^0

答案及解析

1. $w = 26.956$

Quantity A	Quantity B
The digit in the hundredths place of w	The digit in the hundreds place of $4w$

A. Quantity A is greater.

B. Quantity B is greater.

C. The two quantities are equal.

D. The relationship cannot be determined from the information given.

解题思路

题干让我们比较 w 百分位对应的数值与 $4w$ 百位对应的数值的大小关系。

$w = 26.956$，百分位对应的数值为 5，

$4w = 107.824$，百位对应的数值为 1，

$5 > 1$，故数量 A 更大。

答案　A

2. The two-digit positive integer m is equal to twice the sum of its digit.

Quantity A	Quantity B
m	20

A. Quantity A is greater.

B. Quantity B is greater.

C. The two quantities are equal.

D. The relationship cannot be determined from the information given.

题干翻译

两位数的正整数 m 等于它对应的各个数位的数字之和的两倍。

解题思路

设 $m = xy$（x，y 为数位上的数值），
$m = 10x + y$。
根据题干条件可知 $10x + y = 2(x + y)$，
因此 $8x = y$。
数位上的数值的取值范围为 $0 \sim 9$，且首位不能为 0，
所以，$x = 1$，$y = 8$，$m = 18 < 20$，
故数量 B 更大。

答案 B

3. The speed of light is 3×10^8 meters per second, rounded to the nearest 10^8 meters per second. A "light-hour" is the distance that light travels in an hour.

Quantity A	Quantity B
The number of kilometers in a light-hour	10^{10}

A. Quantity A is greater.
B. Quantity B is greater.
C. The two quantities are equal.
D. The relationship cannot be determined from the information given.

题干翻译

光速四舍五入到 10^8 的结果为 3×10^8 米/秒，"光小时" 是光在一小时内传播的距离。

解题思路

本题让我们比较的是以千米为单位，光小时对应的数值与 10^{10} 的大小关系。
距离 = 速度 × 时间，
光小时 = $3 \times 10^8 \times 3600 = 1.08 \times 10^{12}$m。
1m = 10^{-3}km，
因此，光小时 = 1.08×10^{12}m = $1.08 \times 10^{12} \times 10^{-3}$km = 1.08×10^9km。
$1.08 \times 10^9 < 10^{10}$，
故数量 B 更大。

答案 B

4. A certain brand of chewing gum can be purchased in only two ways：
1 piece at a time at 7 cents a piece
5 pieces at a time at 30 cents for the 5 pieces

Quantity A	Quantity B
The greatest number of pieces of the gum that can be purchased with 87 cents	The greatest number of pieces of the gum that can be purchased with 82 cents

A. Quantity A is greater.

B. Quantity B is greater.

C. The two quantities are equal.

D. The relationship cannot be determined from the information given.

题干翻译

某种品牌的口香糖只能通过两种方式购买：

1 条 7 美分

5 条 30 美分

解题思路

本题让我们比较的是 87 美分能买的口香糖的最大可能数量和 82 美分能买的口香糖的最大可能数量。显然，每次买 5 条更实惠，因此要买尽可能多的口香糖，应该尽可能多地采用每次买 5 条的方式。

我们先算一下 87 美分能买的口香糖的最大可能数量：

$87 \div 30 = 2 \cdots\cdots 27$，因此 87 美分最多买 2 次 5 条，即 10 条，剩下的 27 美分再单买口香糖。

$27 \div 7 = 3 \cdots\cdots 6$，因此 27 美分能买的口香糖数量为 3 条。

综上，87 美分最多买 $10 + 3 = 13$ 条口香糖。

我们再算一下 82 美分能买的口香糖最大可能数量：

$82 \div 30 = 2 \cdots\cdots 22$，因此 82 美分最多买 2 次 5 条，即 10 条，剩下的 22 美分再单买口香糖。

$22 \div 7 = 3 \cdots\cdots 1$，因此 27 美分能买的口香糖数量为 3 条。

综上，82 美分最多买 $10 + 3 = 13$ 条口香糖。

数量 A = 数量 B。

答案 C

5. $10^2 (0.00839) =$

A. 8.39×10^{-4}
B. 8.39×10^{-3}
C. 8.39×10^{-2}
D. 8.39×10^{-1}
E. 8.39×10^0

解题思路

$0.00839 = 8.39 \times 10^{-3}$，

$10^2 (0.00839) = 10^2 \times 8.39 \times 10^{-3} = 8.39 \times 10^{2-3} = 8.39 \times 10^{-1}$。

答案 D

1.3 指数和根

基本词汇

exponent 指数　　　　power 幂，次方　　　　base 底数　　　　root 根

1.3.1 指数

概念

指数是用来表示一个数字与自身重复相乘。

例如：$2^4 = (2)(2)(2)(2) = 16$。

在2^4这个表达式中，2 称为底数，4 称为指数，我们把2^4称为 2 的四次方。当指数为 2 的时候，我们称为平方。

例如：5 的平方是$5^2 = (5)(5) = 25$。

负数的偶数次方是正数；负数的奇数次方是负数。

例如：$(-2)^2 = 4$，$(-2)^3 = -8$。

注意，如果没有括号，如$-2^2 (=-4) \neq (-2)^2 (=4)$。

对于任何非零数字 a，$a^0 = 1$，0^0是没有意义的。

对于任何非零数字 a，$a^{-1} = \frac{1}{a}$，$a^{-2} = \frac{1}{a^2}$，$a^{-3} = \frac{1}{a^3}$，以此类推。

注意，$(a)(a^{-1}) = (a)\left(\frac{1}{a}\right) = 1$。

考点1　指数的尾数循环

例题 What is the remainder if 33^{777} is divided by 5?

题干翻译

$33^{777} \div 5$ 的余数是多少？

解题思路

注意，考试的计算器只能帮我们做 +，-，×，÷ 和开平方的计算，是没法帮我们做科学计算的。并且考试的计算器只能显示 8 位及以内的数值。因此当涉及 a^n 的计算，其中 n 特别大（一般大于等于 10 就算特别大了）的时候，我们肯定是无法用计算器得出结果的。因此，我们要知道此类题目肯定是要用到某些解题技巧的。

我们前面说过，5 的倍数的特征是：个位是 0 或 5，而十位、百位及以上数位都是 10 的倍数，都可以被 5 整除。所以，一个数值÷5，余数是多少，取决于这个数的个位数的数值。

这道题问 33^{777}÷5 的余数是多少，那么首先需要知道 33^{777} 的个位数的数值。

以后只要问 a^n 的个位数的数值，都涉及一个知识点：指数的尾数循环。

具体操作如下：

第一步：底数只保留个位

对于本题中的 33^{777} 的底数 33，我们只保留个位上的数值 3。

第二步：通过试数找出个位^{n}的循环周期

对于本题，我们通过试数找出 3^n 的个位数值的循环周期：

3^1 个位数为 3，

3^2 个位数为 9，

3^3 个位数为 7，

3^4 个位数为 1，

3^5 个位数为 3，

……

通过试数我们可以发现，3^n 个位上数值的循环规律为：3，9，7，1。

循环周期长度为 4，即每 4 个数字一循环。

第三步：指数 n÷周期，看余几（余几说明个位^{n}落在周期数字的哪个位置上）。

对于 33^{777}，我们用指数÷周期长度：$777 \div 4 = 194 \cdots\cdots 1$，

余 1 说明 33^{777} 个位对应的是循环规律的第一个位置上的数字。

因此，33^{777} 的个位数为 3，

33^{777}÷5 余数为 3。

指数的尾数循环有两种问法：

① a^n 的个位数是多少？

② a^n÷5 的余数是多少？/a^n÷10 的余数是多少？

对于第二种问法，因为一个数值÷5 的余数与其÷10 的余数，只受个位上数值的影响，因此只看个位数就可以。

> **注意**
>
> GRE 数学经常考查整数的 n 次方个位数的规律，大家可以记一下，以节约考试计算时间。当然，不愿意记的同学，也可以利用乘法口诀推出来。
>
> 个位以 2 结尾的数对应的次方，个位数以 2，4，8，6 循环；
>
> 个位以 3 结尾的数对应的次方，个位数以 3，9，7，1 循环；
>
> 个位以 4 结尾的数对应的次方，个位数以 4，6 循环；
>
> 个位以 6 结尾的数对应的次方，个位数以 6 循环；
>
> 个位以 7 结尾的数对应的次方，个位数以 7，9，3，1 循环；
>
> 个位以 8 结尾的数对应的次方，个位数以 8，4，2，6 循环；
>
> 个位以 9 结尾的数对应的次方，个位数以 9，1 循环。

考点2 a^n（n 特别大）除以非5、非10 的数字的余数问题

例题 When 3^{64} is divided by 8, what is the remainder?

题干翻译

$3^{64} \div 8$ 的余数是多少？

解题思路

我们前面已经提到过考试中的计算器无法帮我们进行科学计算，因此此类题我们需要知道它对应的解题技巧。

与前面讨论的 $a^n \div 5$ 的余数或 $a^n \div 10$ 的余数的情况不同，对于 $a^n \div$ 其他数字的余数，我们不能只看 a^n 的个位循环规律，因为 a^n 除以其他数字时，a^n 的十位、百位或其他数位上的数值的变化可能会影响余数。那么此类题我们应该如何操作呢？

第一步：通过试数（$n = 1, 2, 3, \cdots$）找出 a^n 对应的具体数值

比如本题中，我们应该找到 3^n 对应的具体数值：

$n = 1$，$3^1 = 3$，

$n = 2$，$3^2 = 9$，

$n = 3$，$3^3 = 27$，

$n = 4$，$3^4 = 81$，

……

第二步：用 a^n 对应的具体数值除以除数，找出余数的循环规律

比如本题中，我们应该用 3^n 对应的具体数值除以8，找出余数的循环：

$3^1 \div 8$，余3，

$3^2 \div 8$，余1，

$3^3 \div 8$，余3，

$3^4 \div 8$，余1，

……

通过试数我们可以发现，$3^n \div 8$ 的余数循环规律为：3，1，

循环周期长度为2，即每两个数字一循环。

第三步：用指数 n 除以周期长度，余几就说明落在周期的第几个数字上，找到对应的数字就是我们的答案

对于本题，我们用指数 $n \div$ 周期长度看余数：$64 \div 2 = 32$，是整除，整除说明周期刚好循环完，即落在周期的最后一个数字上，我们找到这个数字为1。

因此，$3^{64} \div 8$ 的余数为1。

例题 01.

If n is a positive integer, which of the following CANNOT be the units digit of $2^n - 1$?
A. 1　　　　　B. 3　　　　　C. 5　　　　　D. 7　　　　　E. 9

题干翻译
如果 n 是正整数，下列哪一项不可能是 $2^n - 1$ 的个位数位对应的数字？

解题思路
2^n 个位数的循环规律是 2，4，8，6，$2^n - 1$ 个位数为 1，3，7，5，所以不可能是 9。

答案　E

例题 02.

What is the remainder when 9^{185} is divided by 10?

题干翻译
9^{185} 除以 10 的余数是多少？

解题思路
对于 a^n 除以 10 的余数问题，我们需要知道 a^n 个位上数值的循环周期，因此想求 9^{185} 除以 10 的余数，我们要知道 9^{185} 个位上对应的数值。
9^n 个位上对应的数值循环规律是：9，1，9，1，循环周期长度是 2。
指数 $n \div$ 周期余几：$185 \div 2$ 余 1，说明 9^{185} 落在第 1 个位置，个位数就是 9，用个位上对应的数字除以 10，算出余数：$9 \div 10$ 的余数是 9。

答案　9

例题 03.

What is the remainder when 8^{43} is divided by 7?

题干翻译
8^{43} 除以 7 的余数是多少？

解题思路
对于 $a^n \div$ 其他数字的余数问题，我们通过试数找出余数的规律来确定余数。
$8 \div 7$ 余 1，
$8^2 \div 7$ 余 1，
$8^3 \div 7$ 余 1，
……
因此，我们可以推出 $8^{43} \div 7$ 余 1。

答案　1

练习

1. What is the product of the units digit of 7^{34} and the units digit of 6^{30}?

2. What is the remainder when 3^{35} is divided by 5?

3. What is the remainder when 2^{5550} is divided by 7?
   A. 6   B. 4   C. 3   D. 2   E. 1

答案及解析

1. What is the product of the units digit of 7^{34} and the units digit of 6^{30}?

题干翻译

7^{34}个位上的数值和6^{30}个位上的数值的乘积是多少?

解题思路

7^n个位上数值的循环规律为：7，9，3，1 这4个数字，$34 \div 4$ 余2，说明7^{34}的个位落在周期中的第二个数字上，即7^{34}个位上对应的数字为9。

6^n个位上数值的循环规律为：6 这1个数字，因此6^{30}个位上对应的数字为6。

$9 \times 6 = 54$。

答案　54

2. What is the remainder when 3^{35} is divided by 5?

题干翻译

3^{35}除以5的余数是多少?

解题思路

$a^n \div 5$ 的余数问题看a^n个位上数值的循环规律：

3^n个位上对应的数值循环规律为：3，9，7，1 这4个数字。

$35 \div 4 = 8 \cdots\cdots 3$，说明$3^{35}$个位上对应的数值为周期的第3个数字。

即3^{35}个位上对应的数字为7，

$7 \div 5$ 余 2。

答案 2

3. What is the remainder when 2^{5550} is divided by 7?

A. 6　　　　B. 4　　　　C. 3　　　　D. 2　　　　E. 1

题干翻译

2^{5550} 除以 7 的余数是多少？

解题思路

$a^n \div 5$ 和 10 以外的其他数字的余数问题直接试数，找余数规律：

$2^1 \div 7$ 余 2，

$2^2 \div 7$ 余 4，

$2^3 \div 7$ 余 1，

$2^4 \div 7$ 余 2，

$2^5 \div 7$ 余 4。

因此，$2^n \div 7$ 的余数循环规律为 2，4，1 这 3 个数字，周期长度为 3。

$5550 \div 3 = 1850$，是整除，说明 $2^{5550} \div 7$ 的余数落在周期的最后一个数字上，

即 $2^{5550} \div 7$ 的余数为 1。

答案 E

1.3.2 根

概念

非零数字 n 的平方根 r 指的是 r 能够使得 $r^2 = n$。

例如，3 是 9 的平方根，因为 $3^2 = 9$。9 的另外一个平方根是 -3，因为 $(-3)^2 = 9$。

所有的正数都有 2 个平方根，一个正的，一个负的。

0 的唯一的平方根为 0。

符号 $\sqrt{n}$ 表达的是非负数 n 的平方根。

在实数定义域下，负数的平方根没有意义。

考点

平方根的运算法则：

$(\sqrt{a})^2 = a$，

$\sqrt{a^2} = |a|$，

$\sqrt{a}\sqrt{b} = \sqrt{ab}$,

$\dfrac{\sqrt{a}}{\sqrt{b}} = \sqrt{\dfrac{a}{b}}$。

例如：

$(\sqrt{2})^2 = 2$,

$\sqrt{2^2} = 2$,

$\sqrt{2}\sqrt{3} = \sqrt{6}$,

$\dfrac{\sqrt{2}}{\sqrt{3}} = \sqrt{\dfrac{2}{3}}$。

$\sqrt{n}$ 或者 $\sqrt[2]{n}$ 表示 n 的平方根，$\sqrt[3]{n}$ 表示 n 的立方根，$\sqrt[4]{n}$ 表示 n 的四次方根，以此类推。

注意，奇数次方根与偶数次方根在实数体系下有明显的区别：

- 对于所有的数值 n，有且仅有 1 个奇数次方根。例如，$\sqrt[3]{27} = 3$。
- 对于每个正数 n，有两个偶数次方根；负数没有偶数次方根。例如，4 有 2 个四次方根：$\sqrt[4]{4}$ 和 $-\sqrt[4]{4}$，但是 -4 没有四次方根。

常用数值的平方根：

$\sqrt{121} = 11$	$\sqrt[3]{8} = 2$
$\sqrt{144} = 12$	$\sqrt[3]{27} = 3$
$\sqrt{169} = 13$	$\sqrt[3]{64} = 4$
$\sqrt{196} = 14$	$\sqrt[3]{125} = 5$
$\sqrt{225} = 15$	$\sqrt[3]{216} = 6$
$\sqrt{625} = 25$	

If $x = \sqrt{10}$, what is the value of $(5\sqrt{10} + x)^2$?

题干翻译

如果 $x = \sqrt{10}$，$(5\sqrt{10} + x)^2$ 的值是多少？

解题思路

$x = \sqrt{10}$,

$5\sqrt{10} + x = 6\sqrt{10}$,

$(6\sqrt{10})^2 = 360$。

答案 360

m and n are integers.

Quantity A	Quantity B
$(\sqrt{10^{2m}})(\sqrt{10^{2n}})$	10^{mn}

A. Quantity A is greater.

B. Quantity B is greater.

C. The two quantities are equal.

D. The relationship cannot be determined from the information given.

题干翻译

m 和 n 为整数。

解题思路

本题让我们比较 $(\sqrt{10^{2m}})(\sqrt{10^{2n}})$ 与 10^{mn} 的大小关系。

数量 A $= (\sqrt{10^{2m}})(\sqrt{10^{2n}}) = 10^m \times 10^n = 10^{m+n}$。

由于题干信息只告诉了我们 m 和 n 为整数，因此我们无法确定 $m+n$ 和 mn 的大小关系，即 $(\sqrt{10^{2m}})(\sqrt{10^{2n}})$ 与 10^{mn} 的大小关系不确定。

答案 D

1.4 实数

基本词汇

real number 实数 absolute value 绝对值
real number line 实数轴 triangle inequality 三角不等式

概念

实数是有理数和无理数的总称，包含整数、分数和小数。

每一个实数在数轴上都有与之相对应的点。在数轴上，0左边的所有数都是负数，0右边的所有数都是正数。0既不是正数也不是负数。

数轴上如果x在y的左侧，那么实数x小于实数y，即$x<y$。数轴上如果x在y的右侧，那么实数x大于实数y，即$x>y$。

在数轴上数字x与0点之间的距离被称为绝对值x，写为$|x|$。因此，$|2|=2$，$|-2|=2$，因为2和-2与0之间的距离都是2。注意，正数的绝对值等于它本身，负数的绝对值等于它的相反数，0的绝对值等于0。

例如：

$|3|=3$，

$|-3|=3$，

$|0|=0$。

考点

如果a，b，c是实数，则

① $a+b=b+a$，$ab=ba$。

例如：$1+2=2+1=3$，$2\times5=5\times2=10$。

② $(a+b)+c=a+(b+c)$，$(ab)c=a(bc)$。

例如：$(1+2)+3=1+(2+3)=6$，$(2\times5)\times3=5\times(2\times3)=30$。

③ $a(b+c)=ab+ac$。

例如：$5(2+3)=5\times2+5\times3=25$。

④ $a+0=a$,$(a)(0)=0$,$(a)(1)=a$。
⑤ 如果 $ab=0$,则 $a=0$ 或 $b=0$ 或 $a=b=0$。
⑥ 0 不能作为除数。
⑦ 如果 a 和 b 都是正数,那么 $a+b$ 和 ab 都是正数。
⑧ 如果 a 和 b 都是负数,那么 $a+b$ 是负数,ab 是正数。
⑨ 如果 a 是正数,b 是负数,那么 ab 是负数。
⑩ 三角不等式:$|a+b| \leq |a|+|b|$。
⑪ $|a||b|=|ab|$。
⑫ 如果 $a>1$,那么 $a^2>a$;如果 $0<b<1$,那么 $b^2<b$。

If $|z| \leq 1$, which of the following statements must be true? Indicate all such statements.

A. $z^2 \leq 1$ B. $z^2 \leq z$ C. $z^3 \leq z$

题干翻译

如果 $|z| \leq 1$,下列哪项一定正确?

解题思路

根据题干条件 $|z| \leq 1$ 可知,$-1 \leq z \leq 1$,

将 $|z| \leq 1$ 两边同时平方,可知 $z^2 \leq 1$,A 选项正确。

当 z 为负数时,$z^2 > 0 > z$,B 选项错误。

当 $-1 < z < 0$ 时,比如 $z=-0.1$,$z^3 > z$,C 选项错误。

答案 A

1.5 比例/比率

基本词汇

ratio 比率　　　　proportion 比例　　　　cross multiplication 交叉相乘

概念

比率是用来表示一个数量与另外一个数量的相对大小关系，通常表示成分数形式。在这个分数形式中，第一个数字为分子部分，第二个数字为分母部分。例如，x 比 y 可以写成 $\frac{x}{y}$，x to y 或 $x:y$。举例：如果篮子里有 2 个苹果，3 个橘子，那么我们可以说苹果与橘子的数量比为 $\frac{2}{3}$ 或 2 to 3 或 2:3。

比率也可以是 3 个及以上的数字之间的大小关系。例如，x 比 y 比 z，可以写成 x to y to z 或 $x:y:z$。举例：如果篮子里有 20 个苹果，30 个橘子，40 个梨，那么我们可以说三者的数量比为 20:30:40。比率可以写成最简形式，上述比率的最简形式为 2:3:4。

比例是关于两个比率的等式，例如，$\frac{20}{30} = \frac{2}{3}$。要解一个关于比率的问题，我们可以写一个比例，然后再交叉相乘。

例题 01.

To find a number x so that the ratio of x to 25 is the same as the ratio of 3 to 5.

题干翻译

找出 x 的数值，使 x 与 25 的比率等于 3 与 5 的比率。

解题思路

$\frac{x}{25} = \frac{3}{5}$，

交叉相乘可得 $5x = 3 \times 25$，

$x = 15$。

答案　15

The perimeters of square region S and rectangular region R are equal. If the sides of R are in the ratio 2∶3, what is the ratio of the area of region R to the area of region S?

A. 25∶16　　　　　B. 24∶25　　　　　C. 5∶6　　　　　D. 4∶5　　　　　E. 4∶9

题干翻译

正方形区域 S 和矩形区域 R 的周长相等。如果 R 的边长比率为 2∶3，那么 R 区域的面积与 S 区域的面积之比是多少？

解题思路

根据题干条件可知，矩形区域 R 的边长比率为 2∶3，设 R 的长度和宽度分别为 $2x$ 和 $3x$。

根据题干条件，正方形区域 S 和矩形区域 R 的周长相等，我们可以算出正方形 S 的边长：

正方形 S 的边长 $= \dfrac{\text{矩形 R 的周长}}{4} = \dfrac{(2x+3x)\times 2}{4} = 2.5x$，

矩形 R 的面积：正方形 S 的面积 $= (2x \times 3x):(2.5x \times 2.5x) = 6x^2:6.25x^2 = 24:25$。

答案 B

练习

1. If x and y are positive numbers and the ratio of x to y is 5 to 4, which of the following ratios must be equal to 6 to 5?

A. $2x$ to $3y$　　　B. $12x$ to $10y$　　　C. $24x$ to $25y$　　　D. $x+1$ to $y+1$　　　E. $x+6$ to $y+5$

2. Some standard paper sizes

Type	Width (millimeters)	Length (millimeters)
A_0	841	1,189
A_1	594	841
A_2	420	594

Quantity A

The ratio of the area of a sheet of A_0 paper to the area of a sheet of A_1 paper

Quantity B

The ratio of the area of a sheet of A_1 paper to the area of a sheet of A_2 paper

A. Quantity A is greater.

B. Quantity B is greater.

C. The two quantities are equal.

D. The relationship cannot be determined from the information given.

答案及解析

1. If x and y are positive numbers and the ratio of x to y is 5 to 4, which of the following ratios must be equal to 6 to 5?

A. $2x$ to $3y$ B. $12x$ to $10y$ C. $24x$ to $25y$

D. $x+1$ to $y+1$ E. $x+6$ to $y+5$

题干翻译

如果 x 和 y 是正数，且 $x:y=5:4$，以下哪个比值一定等于 $6:5$？

解题思路

根据题干条件可知，$x:y=5:4$。

设 $x=5a$，$y=4a$，

代入各个选项，$2x:3y=10a:12a=5:6$，A 选项不符合要求。

$12x:10y=60a:40a=3:2$，B 选项不符合要求。

$24x:25y=120a:100a=6:5$，C 选项正确。

$(x+1):(y+1)=(5a+1):(4a+1)$，由于 a 为变量，因此这个比值结果不确定，D 选项不符合要求。

$(x+6):(y+5)=(5a+6):(4a+5)$，由于 a 为变量，因此这个比值结果不确定，E 选项不符合要求。

答案 C

2. Some standard paper sizes

Type	Width (millimeters)	Length (millimeters)
A_0	841	1,189
A_1	594	841
A_2	420	594

Quantity A	Quantity B
The ratio of the area of a sheet of A_0 paper to the area of a sheet of A_1 paper	The ratio of the area of a sheet of A_1 paper to the area of a sheet of A_2 paper

A. Quantity A is greater.

B. Quantity B is greater.

C. The two quantities are equal.

D. The relationship cannot be determined from the information given.

解题思路

本题让我们比较A_0的面积：A_1的面积与A_1的面积：A_2的面积的大小关系。

如果分别用计算器算出"A_0的面积：A_1的面积"与"A_1的面积：A_2的面积"两者的数值，由于计算器的结果只能显示8个数位，所以很难看出数量A和数量B的大小关系。

本题建议思路：

数量 $A = \dfrac{841 \times 1189}{594 \times 841}$，

数量 $B = \dfrac{594 \times 841}{420 \times 594}$。

我们基于上述表达，算出$\dfrac{数量 A}{数量 B}$的值，再拿这个值与1进行大小比较即可。

$\dfrac{数量 A}{数量 B} = \dfrac{1189 \times 420}{841 \times 594}$，

用计算器按出$\dfrac{数量 A}{数量 B}$的结果<1，

因此，数量 $A <$ 数量 B。

答案　B

1.6 百分比

基本词汇

percent 百分比

概念

百分比指的是一个数是另一个数的百分之几，经常被用来表示把整体当成 100 份之后，所代表的份数。

例如：

1 percent 表示 100 份当中的 1 份，可以写成 $\frac{1}{100}$。

50 percent 表示 100 份当中的 50 份，可以写成 $\frac{50}{100}$ 或 $\frac{1}{2}$。

注意，"在总体中所占的部分"一般写成分子，"总体"一般写成分母。我们通常用%这个符号来表示百分比。百分数也可以用分数和小数来表示。

例如：

$1\% = \frac{1}{100} = 0.01$，

$50\% = \frac{50}{100} = 0.5$。

注意，不要混淆 0.01 与 0.01%。百分比符号是有意义的，$0.01 = 1\%$，但 $0.01\% = \frac{0.01}{100} = 0.0001$。

关于百分比问题在 GRE 中如何考查，我们首先需要区分两种表达形式：

increase/decrease by 增加/下降了……

increase/decrease to 增加/下降到……

如果说小明上个月的消费为 x 元，本月小明的消费与上月相比 decrease by 20%，是在告诉我们小明的消费相比于上个月减少了 20%，因此小明本月的消费为 $(1-20\%)x = 80\%x$ 元。

如果说小明上个月的消费为 x 元，本月小明的消费与上月相比 decrease to 20%，是在告诉我们小明的消费相比于上个月减少到 20%，因此小明本月的消费为 $20\%x$ 元。

大家读题的时候务必看清楚，题干中的描述用的是 by 还是 to：by 表示的是变化的程度，to 后面跟的是最终的结果。

下面我们一起来看一下百分比问题的具体考法。

考点1　已知两个数值，求变化的百分比

A 比 B 大/增加了多少百分比 $= \dfrac{大-小}{小} \times 100\%$。

B 比 A 小/减少了多少百分比 $= \dfrac{大-小}{大} \times 100\%$。

比如：

① 已知小明7月工资为800美元，8月工资为1000美元，请问从7月到8月小明的工资增长了百分之几？

小明7月到8月工资增长百分比 $= \dfrac{大-小}{小} \times 100\% = \dfrac{1000-800}{800} \times 100\% = 25\%$

② 已知小明7月工资为1000美元，8月工资为800美元，请问从7月到8月小明的工资减少了百分之几？

小明7月到8月工资减少百分比 $= \dfrac{大-小}{大} \times 100\% = \dfrac{1000-800}{1000} \times 100\% = 20\%$。

考点2　已知变化百分比，求具体数值

A 比 B 大 $x\%$，则 $A = B \times (1+x\%)$，

B 比 A 小 $y\%$，则 $B = A \times (1-y\%)$。

考点3　关于百分比、比例的读题问题

我们首先需要区分以下两种表达：

① 数学好的女生在我们班女生中占了60%；

② 数学好的女生在我们班学生中占了60%。

这两个表达并不一样：第一个表达中，60%是针对"我们班女生的数量"来说的，而第二个表达中，60%是针对"我们班学生的数量"来说的。

读题技巧：在英文表达中，如何知道百分比和比率是针对谁而言的呢？关键看的是 percent of 和 fraction of 后面紧跟的对象。

关于百分比、比例的读题问题，主要考查以下几种表达方式：

1. percent/fraction

What percent of A is B?　　B 是 A 的百分之几？

这个表达中，A 是分母，B 是分子，what percent $= \dfrac{B}{A}$。

2. ratio

the ratio of A to B　　A 与 B 的比值

这个表达永远是：to 前的数值 : to 后的数值。

3. times

(1) *A* is two times as many as *B*.
(2) (There is) two times as many *A* as *B*.

$\left.\begin{array}{l}\\\\\end{array}\right\}$ *A* 是 *B* 的 2 倍，即 $A = 2B$

题干用 times 表示倍数时，可以直接翻译为"前者为后者的几倍"。

If *x* and *y* are positive and *x* is 43 percent greater than *y*, then *y* must be what percent less than *x*? Give your answer to the nearest whole percent.

_____ percent

题干翻译

如果 *x* 和 *y* 都是正数，且 *x* 比 *y* 大 43%，那么 *y* 一定比 *x* 小百分之几？

给出你的答案，精确到最接近的整百分数。

解题思路

根据题干条件"*x* 比 *y* 大 43%"可知，$x = (1+43\%)y = 1.43y$。

$$y \text{ 比 } x \text{ 小百分之几} = \frac{\text{大} - \text{小}}{\text{大}} \times 100\%$$

$$= \frac{x-y}{x} \times 100\%$$

$$= \frac{1.43y - y}{1.43y} \times 100\%$$

$$= \frac{0.43y}{1.43y} \times 100\%$$

$$\approx 30\%$$

答案 30

Today a certain machine is worth 20 percent less than it was worth a year ago, and it is worth *x* percent less than it was worth two years ago. A year ago the machine was worth 20 percent less than it was worth two years ago.

Quantity A   Quantity B
   x   40

A. Quantity A is greater.
B. Quantity B is greater.

C. The two quantities are equal.
D. The relationship cannot be determined from the information given.

题干翻译

今年某台机器的价格比一年前少了 20%，相比于两年前少了 $x\%$。一年前，这台机器的价格比两年前低 20%。

解题思路

本题让我们比较 x 和 40 的大小关系。

设前年机器价格为 a，

根据题干条件"今年某台机器的价格比一年前少了 20%，且一年前，这台机器的价格比两年前低 20%"可知，今年的机器价格 $=(1-20\%)(1-20\%)a=64\%a$。

根据题干条件"今年某台机器的价格比一年前少了 20%，相比于两年前少了 $x\%$"可知，今年的机器价格 $=(1-x\%)a$。

$64\%a=(1-x\%)a$，

$x\%=36\%$，

$x=36<40$，

故数量 B 更大。

答案 B

In a certain state, 60 percent of the eligible voters are registered to vote and 48 percent of the eligible voters in that state voted in a recent election. If only registered voters are permitted to vote, what percent of the registered voters in the state voted in the election?

A. 30% B. 50% C. 60% D. 70% E. 80%

题干翻译

在某个州，60% 的合格选民注册投票，并且该州 48% 的合格选民在最近的选举中投票。如果只允许做过注册的选民投票，那么该州有百分之多少的做过注册的选民在选举中投票？

解题思路

阅读题干，eligible voters 中有 60% 的人注册了，有 48% 的人在最近的选举中投票了（注意，60% 和 48% 中 percent of 后面接的都是 eligible voters，因此这两个百分比是针对 eligible voters 说的）。只有注册过的选民才被允许投票，问在 registered voters 中参与了投票的选民占比（注意，题干中的百分比是指在 the registered voters 中的占比，因为 percent of 紧跟的是 the registered voters）。

即 $\dfrac{\text{the registered}}{\text{the eligible}}=60\%$，$\dfrac{\text{the voted}}{\text{the eligible}}=48\%$，

则 $\dfrac{\text{the voted}}{\text{the registered}}=\dfrac{48\%}{60\%}=80\%$。

答案 E

例题 04.

In a certain building in a business district, a newspaper service delivers 104 papers every day to its customers. Twice as many customers have 3 papers delivered each day as have one paper delivered, and three times as many customers have 2 papers delivered each day as have one paper delivered. If no customer has more than 3 papers delivered, then the number of customers who have 2 papers delivered each day is _____.

A. 16   B. 18   C. 20   D. 22   E. 24

题干翻译

在商业区的某栋大楼里,一家报纸服务公司每天向其客户发送 104 份报纸。每天得到 3 份报纸的客户是得到 1 份报纸的两倍,每天得到 2 份报纸的客户是得到 1 份报纸的三倍。如果没有客户得到 3 份以上的报纸,那么每天得到 2 份报纸的客户数量是多少?

解题思路

设每天得到 1 份报纸的客户为 x 个,那么根据题干条件可知,得到 3 份报纸的客户有 $2x$ 个,得到 2 份报纸的客户有 $3x$ 个。

报纸总数 $= x + 3 \times 2x + 2 \times 3x = 104$,

$13x = 104$,

$x = 8$。

因此,得到 2 份报纸的客户数量 $3x = 3 \times 8 = 24$。

答案 E

练 习

1. A merchant bought a certain item from a manufacturer at a wholesale price of \$15.00. The manufacturer recommended a price, called the suggested retail price, at which the merchant could sell the item. The merchant actually sold the item for an amount that was 20 percent greater than the wholesale price and 20 percent less than the suggested retail price. What was the suggested retail price of the item?

A. \$18.00   B. \$21.00   C. \$21.60   D. \$22.50   E. \$23.40

2.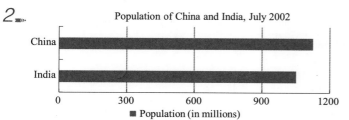

According to the graph, in July 2002, the population of China was approximately what percent greater than the population of India?

A. 2%   B. 7%   C. 17%   D. 22%   E. 27%

3. Let x, y, and z be positive numbers. If x is 12.5 percent of y and x is 3.125 percent of z, then x is what percent of $y + z$?

(Enter your answer as an integer or a decimal in the answer box. Backspace to erase.)

_____ %

4. | Quantity A | Quantity B |
|---|---|
| 32 percent of 15 | 5 |

A. Quantity A is greater.
B. Quantity B is greater.
C. The two quantities are equal.
D. The relationship cannot be determined from the information given.

5. SOURCES OF FINANCIAL SUPPORT FOR A NONPROFIT ORGANIZATION

Source	Percent of Financial Support
Corporations	37%
Federal/state governments	33%
Individuals	22%
Others	8%

The table shows the percent distribution of financial support for a nonprofit organization, by source. The amount of financial support from individuals is approximately what percent less than the amount from corporations?

A. 24%   B. 35%   C. 41%   D. 69%   E. 88%

6. Of all the children living in Country M in 1989, 24 percent had access to a computer at home and 18 percent used a computer at home. What percent of children who had access to a computer at home did not use a computer at home?

A. 6% B. 25% C. 33% D. 38% E. 82%

7. Bucket Q currently contains 5 times as much water as bucket R contains. If 10 liters of water will be transferred from Q to R, then Q will contain 2 times as much water as R will contain. How many liters of water does Q currently contain?

A. 10 B. 20 C. 30 D. 40 E. 50

答案及解析

1. A merchant bought a certain item from a manufacturer at a wholesale price of $15.00. The manufacturer recommended a price, called the suggested retail price, at which the merchant could sell the item. The merchant actually sold the item for an amount that was 20 percent greater than the wholesale price and 20 percent less than the suggested retail price. What was the suggested retail price of the item?

A. $18.00 B. $21.00 C. $21.60 D. $22.50 E. $23.40

题干翻译

一个商人以 15.00 美元的批发价从制造商那里购买了某种商品。制造商推荐了一个价格，称为建议零售价，在这个价格下，商家可以出售该商品。商家实际上以比批发价高 20%，比建议零售价低 20% 的价格出售了该商品。这件商品的建议零售价是多少？

解题思路

设建议零售价（suggested retail price）为 x。

根据条件"商家实际上以比批发价高 20% 的价格出售了该商品"可知，

实际售价 = 批发价 × (1 + 20%) = 15 × 1.2 = 18。

根据条件"商家实际上以比建议零售价低 20% 的价格出售了该商品"可知，

实际售价 = 建议零售价 × (1 − 20%) = 0.8x，

18 = 0.8x，

x = 22.5。

答案 D

2.

Population of China and India, July 2002

China
India

0   300   600   900   1200
■ Population (in millions)

According to the graph, in July 2002, the population of China was approximately what percent greater than the population of India?

A. 2% B. 7% C. 17% D. 22% E. 27%

题干翻译

根据图表，2002 年 7 月，中国人口大约比印度人口多百分之几？

解题思路

根据图表可知，中国人数大约为 1125 million，印度人数大约为 1050 million。

中国比印度人口多的百分比 = $\dfrac{\text{中国人口} - \text{印度人口}}{\text{印度人口}} = \dfrac{1125 - 1050}{1050} \approx 7\%$。

答案 B

3. Let x, y, and z be positive numbers. If x is 12.5 percent of y and x is 3.125 percent of z, then x is what percent of $y + z$?

(Enter your answer as an integer or a decimal in the answer box. Backspace to erase.)

_____%

题干翻译

x, y, z 为正数。如果 x 是 y 的 12.5%，x 是 z 的 3.125%，那么 x 是 $y + z$ 的百分之几？

解题思路

根据题干条件"x 是 y 的 12.5%"可知，

$x = 12.5\%\, y$,

$y = 8x$。

根据题干条件"x 是 z 的 3.125%"可知，

$x = 3.125\%\, z$,

$z = 32x$。

因此，x 是 $y + z$ 的百分之几 = $\dfrac{x}{y+z} = \dfrac{x}{40x} = 2.5\%$。

答案 2.5

4.   Quantity A   Quantity B
    32 percent of 15   5

   A. Quantity A is greater.
   B. Quantity B is greater.
   C. The two quantities are equal.
   D. The relationship cannot be determined from the information given.

解题思路
数量 A：32 percent of 15 $= 32\% \times 15 = 4.8 < 5$，
数量 B 更大。

答案   B

5. **SOURCES OF FINANCIAL SUPPORT FOR A NONPROFIT ORGANIZATION**

Source	Percent of Financial Support
Corporations	37%
Federal/state governments	33%
Individuals	22%
Others	8%

The table shows the percent distribution of financial support for a nonprofit organization, by source. The amount of financial support from individuals is approximately what percent less than the amount from corporations?

  A. 24%   B. 35%   C. 41%   D. 69%   E. 88%

题干翻译
该表格按来源显示了非营利组织财政支持的百分比分布。来自个人的财政支持金额大约比来自公司的金额少百分之几？

解题思路
$\dfrac{公司 - 个人}{公司} = \dfrac{37\% - 22\%}{37\%} \approx 41\%$。

答案   C

6. Of all the children living in Country M in 1989, 24 percent had access to a computer at home and 18 percent used a computer at home. What percent of children who had access to a computer at home did not use a computer at home?

  A. 6%   B. 25%   C. 33%   D. 38%   E. 82%

题干翻译

1989 年，生活在 M 国的所有儿童中，24% 在家里有电脑，18% 在家里使用电脑。在家里有电脑的儿童中，有百分之多少没有在家里使用电脑？

解题思路

生活在 M 国的所有儿童中，家里有电脑但没有使用的儿童百分比 = 24% − 18% = 6%。

"家里有电脑但没有使用的儿童"与"家里有电脑的儿童"的占比为：

$$\frac{\text{家里有电脑但没有使用的儿童在 M 国所有儿童中的占比}}{\text{家里有电脑的儿童在 M 国所有儿童中的占比}} = \frac{6\%}{24\%} = 25\%。$$

答案 B

7. Bucket Q currently contains 5 times as much water as bucket R contains. If 10 liters of water will be transferred from Q to R, then Q will contain 2 times as much water as R will contain. How many liters of water does Q currently contain?

A. 10　　　　　B. 20　　　　　C. 30　　　　　D. 40　　　　　E. 50

题干翻译

Q 桶目前的储水量是 R 桶的 5 倍。如果 10 升水从 Q 桶转移到 R 桶，那么 Q 桶的储水量是 R 桶的 2 倍。Q 桶目前储有多少升水？

解题思路

设 R 桶现在有水 x 升，则 Q 桶中有水 $5x$ 升，

$5x - 10 = 2(x + 10)$，

$x = 10$，

Q 桶当前的储水量 = $5x$ = 50 升。

答案 E

第二章 CHAPTER

代数

代数部分主要考查初中所学的数学内容，包括指数运算，分解和简化代数表达式，函数、方程和不等式等。

代数部分主要考查9个知识点：
（1）代数式运算
（2）指数运算
（3）根号运算
（4）解线性方程
（5）解一元二次方程
（6）解不等式
（7）数列
（8）函数
（9）函数图形

2.1 代数式运算

基本词汇

algebraic expression 代数式
term 项
like terms 同类项
constant term 常数项
coefficient 系数
identity 恒等式
equation 方程
solution 解

概念

代数式是指包含一个或多个变量的表达式。

例如：$3x$，$y+1$，$a^2b+2b^2-b^2+1$，$\dfrac{2}{x+y}$。其中，$3x$ 只有一项，$y+1$ 有两项，$a^2b+2b^2-b^2+1$ 有四项，$\dfrac{2}{x+y}$ 只有一项。

如果一个代数式中的两项变量相同且变量对应的指数也相同，那么这两项被称为同类项。没有变量的项被称为常数项。变量前面的数字被称为系数。

在 $a^2b+2b^2-b^2+1$ 中，$2b^2$ 与 $-b^2$ 为同类项，1 为常数项，2 为 $2b^2$ 这一项的系数。

同类项可以进行合并，用同类项前面的系数进行加减即可。

例如：$2x+3x=5x$，
$$a^2b+2b^2-b^2+1=a^2b+b^2+1。$$

所有变量都恒相等的等式被称为恒等式，常见的恒等式有以下几个。

（1）完全平方公式：
$$(a+b)^2=a^2+2ab+b^2，(a-b)^2=a^2-2ab+b^2。$$

（2）完全立方公式：
$$(a+b)^3=a^3+3a^2b+3ab^2+b^3，(a-b)^3=a^3-3a^2b+3ab^2-b^3。$$

（3）平方差公式：
$$a^2-b^2=(a-b)(a+b)。$$

两个代数式形成一个等式且只有某些变量的值才能满足这个等式，被称为方程。满足这个等式变量所对应的值称为这个方程的解，以下是常见的几种方程类型。

一元一次方程：$x+5=3$。

二元一次方程：$x+3y=5$。

一元二次方程：$3x^2+x-1=0$。

例题 01.

$$x^{-2}=\frac{1}{y+1}$$

Quantity A	Quantity B
x^2	y

A. Quantity A is greater.

B. Quantity B is greater.

C. The two quantities are equal.

D. The relationship cannot be determined from the information given.

解题思路

$x^{-2}=\frac{1}{y+1}$，$\frac{1}{x^2}=\frac{1}{y+1}$，$x^2=y+1>y$。

答案　A

例题 02.

$xy=1$

Quantity A	Quantity B
$(x+y)^2-(x-y)^2$	4

A. Quantity A is greater.

B. Quantity B is greater.

C. The two quantities are equal.

D. The relationship cannot be determined from the information given.

解题思路

数量 A $= (x+y)^2-(x-y)^2$
$= (x^2+2xy+y^2)-(x^2-2xy+y^2)$
$= 4xy$
$= 4$。

数量 A 与数量 B 大小相等。

答案　C

练习

1. $\dfrac{a^2}{b^2} = \dfrac{a}{b}$ ($ab \neq 0$)

Quantity A	Quantity B
a	b

A. Quantity A is greater.

B. Quantity B is greater.

C. The two quantities are equal.

D. The relationship cannot be determined from the information given.

2. $a \neq 0$

Quantity A	Quantity B
$a + 1$	$\dfrac{1}{a} - 1$

A. Quantity A is greater.

B. Quantity B is greater.

C. The two quantities are equal.

D. The relationship cannot be determined from the information given.

3. $x > 0$, $y > 0$

Quantity A	Quantity B
$x + y$	$\sqrt{x^2 + y^2}$

A. Quantity A is greater.

B. Quantity B is greater.

C. The two quantities are equal.

D. The relationship cannot be determined from the information given.

答案及解析

1. $\dfrac{a^2}{b^2} = \dfrac{a}{b}$ ($ab \neq 0$)

Quantity A   Quantity B
  a b

A. Quantity A is greater.

B. Quantity B is greater.

C. The two quantities are equal.

D. The relationship cannot be determined from the information given.

解题思路

$\dfrac{a^2}{b^2} = \dfrac{a}{b}$，

将该分数等式进行交叉相乘可得：$a^2 b = ab^2$，

由于 $ab \neq 0$，等式两边同时除以 ab 可得：$a = b$，

即数量 A 与数量 B 相等。

答案　C

2. $a \neq 0$

Quantity A   Quantity B
 $a + 1$ $\dfrac{1}{a} - 1$

A. Quantity A is greater.

B. Quantity B is greater.

C. The two quantities are equal.

D. The relationship cannot be determined from the information given.

解题思路

试数：

当 $a = 1$ 的时候，数量 A = 2，数量 B = 0，数量 A > 数量 B；

当 $a = -3$ 的时候，数量 A = -2，数量 B = $-\dfrac{4}{3}$，数量 A < 数量 B；

因此数量 A 与数量 B 的大小无法确定。

答案　D

3. $x > 0$, $y > 0$

Quantity A	Quantity B
$x + y$	$\sqrt{x^2 + y^2}$

A. Quantity A is greater.

B. Quantity B is greater.

C. The two quantities are equal.

D. The relationship cannot be determined from the information given.

解题思路

数量 $A = x + y = \sqrt{(x+y)^2} = \sqrt{x^2 + y^2 + 2xy}$,

根据题干条件 $x > 0$, $y > 0$ 可知, $xy > 0$,

因此 $\sqrt{x^2 + y^2 + 2xy} > \sqrt{x^2 + y^2}$,

故数量 A 更大。

答案 A

2.2 指数运算

基本词汇

base 底数　　　　　　　　　　　exponent 指数

概念

在代数式 a^x 中，a 被称为底数，x 被称为指数。

考点1　指数运算法则

以下是常见的指数运算法则，其中 a 和 b 都是非零实数，且 m 和 n 都是整数。

$a^m \times a^n = a^{m+n}$

$a^m \times b^m = (a \times b)^m$

$(a^m)^n = a^{mn}$

$a^m \div a^n = a^{m-n}$

$a^m \div b^m = \left(\dfrac{a}{b}\right)^m$

$a^{-n} = \dfrac{1}{a^n}$

$a^0 = 1$

考点2　指数 a^n 的图像

(1) 如果 $a > 0$，则 $a^n > 0$ 恒成立。

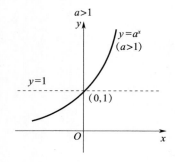
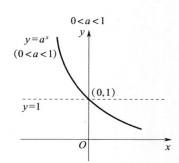

如果 $a>1$，则 a^n 的图像是递增形式（如上面左图）；

如果 $0<a<1$，则 a^n 的图像是递减形式（如上面右图）；

a^n 经过点 (0, 1)。

(2) 如果 $a<0$，则 a^n 的大小取决于指数 n 的奇偶性：

如果指数为偶数，则 $a^n>0$；

如果指数为奇数，则 $a^n<0$。

例题 01.

If $3^x+3^x+3^x=9^{x-2}$, what is the value of x?

$x = $ _____

(Enter your answer as an integer or a decimal in the answer box. Backspace to erase.)

题干翻译

如果 $3^x+3^x+3^x=9^{x-2}$，那么 x 的值是多少？

解题思路

$3^x+3^x+3^x=3\times 3^x=3^{x+1}$，

$9^{x-2}=(3^2)^{x-2}=3^{2x-4}$，

即 $3^{x+1}=3^{2x-4}$，

$x+1=2x-4$，

$x=5$。

答案 5

例题 02.

n is an integer.

Quantity A	Quantity B
$\left(\dfrac{2}{3}\right)^n \left(\dfrac{3}{2}\right)^{-n}$	1

A. Quantity A is greater.

B. Quantity B is greater.

C. The two quantities are equal.

D. The relationship cannot be determined from the information given.

题干翻译

n 为整数。

解题思路

本题让我们比较 $\left(\dfrac{2}{3}\right)^n \left(\dfrac{3}{2}\right)^{-n}$ 与 1 的大小关系。

数量 A $= \left(\dfrac{2}{3}\right)^n \left(\dfrac{3}{2}\right)^{-n} = \left(\dfrac{2}{3}\right)^{2n}$。

注意，题干只告诉我们 n 为整数，但没说正负，所以 $\left(\dfrac{2}{3}\right)^{2n}$ 与 1 的大小关系无法确定。

若 n 为正整数，则 $\left(\dfrac{2}{3}\right)^{2n} < 1$；

若 n 为负整数，则 $\left(\dfrac{2}{3}\right)^{2n} > 1$。

答案　D

例题 03.

x is a negative integer.

Quantity A	Quantity B
$(2x-1)^{2x-1}$	0

A. Quantity A is greater.
B. Quantity B is greater.
C. The two quantities are equal.
D. The relationship cannot be determined from the information given.

题干翻译

x 是一个负整数。

解题思路

$x < 0$，

则 $2x - 1 < 0$，

且 $2x - 1$ 为奇数。

因此，$(2x-1)^{2x-1} < 0$（a^n 底数为负数，指数为奇数，指数运算不改变正负）。

即数量 B 更大。

答案　B

例题 04.

If $x + \dfrac{1}{x} = 5$, then $x^2 + \dfrac{1}{x^2} = $ _____

题干翻译

如果 $x + \dfrac{1}{x} = 5$，那么 $x^2 + \dfrac{1}{x^2} = $ _____

解题思路

将 $x + \dfrac{1}{x} = 5$ 这个等式两边同时进行平方：

$\left(x + \dfrac{1}{x}\right)^2 = 25$，

$x^2 + \dfrac{1}{x^2} + 2 = 25$，

$x^2 + \dfrac{1}{x^2} = 23$。

答案　23

练 习

1. If $8^x = 16$, then $x = $ _____

A. $-\dfrac{1}{2}$ B. $-\dfrac{1}{4}$ C. $\dfrac{1}{2}$ D. $\dfrac{3}{4}$ E. $\dfrac{4}{3}$

2. $x < 0$

Quantity A	Quantity B
$(x^{-1})^{-3}$	x^{-4}

A. Quantity A is greater.

B. Quantity B is greater.

C. The two quantities are equal.

D. The relationship cannot be determined from the information given.

3. x and y are each negative integers.

Quantity A	Quantity B
x^y	0

A. Quantity A is greater.

B. Quantity B is greater.

C. The two quantities are equal.

D. The relationship cannot be determined from the information given.

4. $(x^{n+1})^{n-1} = x^8$ and $x > 1$

Quantity A	Quantity B
The sum of all possible values of n	0

A. Quantity A is greater.

B. Quantity B is greater.

C. The two quantities are equal.

D. The relationship cannot be determined from the information given.

答案及解析

1. If $8^x = 16$, then $x = $ _____

A. $-\dfrac{1}{2}$ B. $-\dfrac{1}{4}$ C. $\dfrac{1}{2}$ D. $\dfrac{3}{4}$ E. $\dfrac{4}{3}$

解题思路

$8^x = (2^3)^x = 2^{3x}$，

$16 = 2^4$，

$3x = 4$，

$x = \dfrac{4}{3}$。

答案 E

2. $x < 0$

Quantity A	Quantity B
$(x^{-1})^{-3}$	x^{-4}

A. Quantity A is greater.

B. Quantity B is greater.

C. The two quantities are equal.

D. The relationship cannot be determined from the information given.

解题思路

数量 A $= (x^{-1})^{-3} = x^3$，根据题干条件 $x < 0$ 可知，$x^3 < 0$（底数为负数，指数为奇数，指数运算不改变正负），

数量 B $= x^{-4} > 0$（a^n 底数为负数，指数为偶数时，$a^n > 0$）。

即数量 A < 数量 B。

答案 B

3. x and y are each negative integers.

Quantity A	Quantity B
x^y	0

A. Quantity A is greater.

B. Quantity B is greater.

C. The two quantities are equal.

D. The relationship cannot be determined from the information given.

题干翻译

x 和 y 都是负整数。

解题思路

本题让我们比较 x^y 与 0 的大小关系。

根据题干条件可知，$x<0$，此时 x^y 与 0 的大小关系由 y 的奇偶性决定。

由于题目信息没有说明 y 的奇偶性，

因此无法确定 x^y 与 0 的大小关系。

答案 D

4. $(x^{n+1})^{n-1} = x^8$ and $x > 1$

Quantity A	Quantity B
The sum of all possible values of n	0

A. Quantity A is greater.

B. Quantity B is greater.

C. The two quantities are equal.

D. The relationship cannot be determined from the information given.

解题思路

本题让我们比较 n 对应的所有可能值的和与 0 的大小关系。

$(x^{n+1})^{n-1} = x^{n^2-1} = x^8$，

$n^2 - 1 = 8$，

$n^2 = 9$，

$n = \pm 3$，

n 所有可能值的和 $= 3 + (-3) = 0$。

故数量 A 与数量 B 相等。

答案 C

2.3 根号运算

考点1 根号运算法则

$\sqrt{a^2} = |a|$ （如果是未知数开根号，结果一定要加绝对值符号）

$\sqrt{a}\sqrt{b} = \sqrt{ab}$

$\dfrac{\sqrt{a}}{\sqrt{b}} = \sqrt{\dfrac{a}{b}}$

注意：$\sqrt{a} \pm \sqrt{b} \neq \sqrt{a \pm b}$

考点2 根号的性质

对于 $\sqrt[m]{a}$ 而言：

如果 m 是偶数，则必须满足 $a \geqslant 0$，且 $\sqrt[m]{a} \geqslant 0$（开偶数次方根的双重非负性：当开偶数次方根时，底数必须大于等于0，且运算结果大于等于0）；

如果 m 是奇数，则 a 可正可负，$\sqrt[m]{a}$ 也可正可负。

例如：$(\pm 2)^2 = 4$，但是 $\sqrt{4} = 2$。

$\sqrt[n]{a^n}$ 的运算结果又如何呢？

当 n 为奇数，$\sqrt[n]{a^n} = a$，

当 n 为偶数，$\sqrt[n]{a^n} = |a|$。

考点3 根号式估值

由于 GRE 数学考试的计算器只能帮我们做开平方根的计算，因此当我们遇到开其他次方根（如开立方根）的情况时，这类题目就涉及估值的问题。

如要估算 $\sqrt[3]{4}$ 的值的范围，我们可以用与底数接近的立方数来估算，如 $1^3 < 4 < 2^3$，则 $1 < \sqrt[3]{4} < 2$。

考点 4　根号运算的化简

对于含有根号的分数式，我们习惯先把分母化为整数。

例如：化简 $\dfrac{\sqrt{21}-\sqrt{6}}{\sqrt{2}}$。

解题思路

$$\dfrac{\sqrt{21}-\sqrt{6}}{\sqrt{2}}=\dfrac{\sqrt{21}-\sqrt{6}}{\sqrt{2}}\times\dfrac{\sqrt{2}}{\sqrt{2}}=\dfrac{\sqrt{42}-2\sqrt{3}}{2}。$$

又如：化简 $\dfrac{\sqrt{2}+1}{\sqrt{2}-1}$。

解题思路

$$\dfrac{\sqrt{2}+1}{\sqrt{2}-1}=\dfrac{\sqrt{2}+1}{\sqrt{2}-1}\times\dfrac{\sqrt{2}+1}{\sqrt{2}+1}=\dfrac{(\sqrt{2}+1)(\sqrt{2}+1)}{(\sqrt{2}-1)(\sqrt{2}+1)}=\dfrac{2+1+2\sqrt{2}}{2-1}=3+2\sqrt{2}。$$

例题 01.

$x\neq 0$

Quantity A	Quantity B
$\dfrac{\sqrt[3]{54x^5}-\sqrt[3]{16x^5}}{x(\sqrt[3]{128x^2}-\sqrt[3]{16x^2})}$	$\dfrac{1}{2}$

A. Quantity A is greater.
B. Quantity B is greater.
C. The two quantities are equal.
D. The relationship cannot be determined from the information given.

解题思路

我们要对数字有敏感性。看这个开三次方根的题，可以考虑对根号下的数字进行分解因数（开三次方，则要看数字里包含的立方数），从而化简表达式。

$54=2\times 3^3$，因此 $\sqrt[3]{54x^5}=3\sqrt[3]{2}\,x^{\frac{5}{3}}$，

$16=2\times 2^3$，因此 $\sqrt[3]{16x^5}=2\sqrt[3]{2}\,x^{\frac{5}{3}}$，

$128=2\times 4^3$，因此 $\sqrt[3]{128x^2}=4\sqrt[3]{2}\,x^{\frac{2}{3}}$，

$16=2\times 2^3$，因此 $\sqrt[3]{16x^2}=2\sqrt[3]{2}\,x^{\frac{2}{3}}$。

$$\dfrac{\sqrt[3]{54x^5}-\sqrt[3]{16x^5}}{x(\sqrt[3]{128x^2}-\sqrt[3]{16x^2})}=\dfrac{3\sqrt[3]{2}\,x^{\frac{5}{3}}-2\sqrt[3]{2}\,x^{\frac{5}{3}}}{x(4\sqrt[3]{2}\,x^{\frac{2}{3}}-2\sqrt[3]{2}\,x^{\frac{2}{3}})}$$

$$= \frac{\sqrt[3]{2}\, x^{\frac{1}{3}}}{x \cdot 2\sqrt[3]{2}\, x^{\frac{2}{3}}}$$

$$= \frac{1}{2} \circ \ (x \cdot x^{\frac{2}{3}} = x^{\frac{5}{3}})$$

注意，如果是开偶数次方根，我们要保证开出来的结果的非负性。

如：$\sqrt{x^2} = |x|$。如果题目中的信息没有告知 x 的正负信息，我们去掉根号后的结果必须带上绝对值。

本题是开奇数次方根，所以不用担心这个问题。

例题 02.

$\sqrt{y^2} = 8$

Quantity A　　　　　　　　**Quantity B**

3^{2y}　　　　　　　　　　　3^{-2y}

A. Quantity A is greater.
B. Quantity B is greater.
C. The two quantities are equal.
D. The relationship cannot be determined from the information given.

解题思路

$\sqrt{y^2} = 8$，

$\sqrt{y^2} = |y| = 8$，

$y = \pm 8$，

当 $y = 8$ 时，$3^{2y} > 3^{-2y}$，

当 $y = -8$ 时，$3^{2y} < 3^{-2y}$，

因此，3^{2y} 与 3^{-2y} 的大小关系无法确定。

答案　D

例题 03.

$0 < x < 1$

Quantity A　　　　　　　　**Quantity B**

$\sqrt{x} + \sqrt{1-x}$　　　　　　　1

A. Quantity A is greater.
B. Quantity B is greater.
C. The two quantities are equal.
D. The relationship cannot be determined from the information given.

解题思路

如果 $0 < a < 1$,则 $a < \sqrt{a}$(举例,如果 $a = 0.01$,则 $\sqrt{a} = 0.1$)。

根据题干条件 $0 < x < 1$ 可知,$0 < 1 - x < 1$。

由于 $0 < a < 1$,$a < \sqrt{a}$,

因此,$\sqrt{x} > x$,$\sqrt{1-x} > 1 - x$,

数量 A $= \sqrt{x} + \sqrt{1-x} > x + 1 - x = 1 =$ 数量 B,

故数量 A 更大。

答案 A

Quantity A	Quantity B
$(\sqrt{999,999,999} + \sqrt{1,000,000,001})^2$	$(\sqrt[3]{999,999,999} + \sqrt[3]{1,000,000,001})^3$

A. Quantity A is greater.
B. Quantity B is greater.
C. The two quantities are equal.
D. The relationship cannot be determined from the information given.

解题思路

本题让我们比较 $(\sqrt{999,999,999} + \sqrt{1,000,000,001})^2$ 与 $(\sqrt[3]{999,999,999} + \sqrt[3]{1,000,000,001})^3$ 的大小关系。

GRE 数学考试的计算器只能做八位数以内以及开平方的计算,所以本题是无法通过计算器来计算的。当计算器无法帮助我们的时候,这类题肯定有相应的技巧,比如当遇到开立方根的计算时,往往可能是考查估值。

与 999999999 和 1000000001 最接近的数字为 10^9,

数量 A $= (\sqrt{999999999} + \sqrt{1000000001})^2$,

$\approx (\sqrt{10^9} + \sqrt{10^9})^2$,

$= (2\sqrt{10^9})^2$,

$= 4 \times 10^9$;

数量 B $= (\sqrt[3]{999999999} + \sqrt[3]{1000000001})^3$,

$\approx (\sqrt[3]{10^9} + \sqrt[3]{10^9})^3$,

$= (2\sqrt[3]{10^9})^3$,

$= 8 \times 10^9 > 4 \times 10^9$;

所以数量 B 更大。

答案 B

例题 05.

Which of the following is equal to $\dfrac{1}{\dfrac{\sqrt{2}+1}{\sqrt{2}-1}}$?

A. $3 - 2\sqrt{2}$ B. $3 + 2\sqrt{2}$ C. $\dfrac{3 - 2\sqrt{2}}{3}$ D. $\dfrac{3 + 2\sqrt{2}}{3}$ E. $\dfrac{1}{3 + \sqrt{2}}$

题干翻译

以下哪一个选项等于 $\dfrac{1}{\dfrac{\sqrt{2}+1}{\sqrt{2}-1}}$?

解题思路

$\dfrac{1}{\dfrac{\sqrt{2}+1}{\sqrt{2}-1}} = \dfrac{\sqrt{2}-1}{\sqrt{2}+1} = \dfrac{\sqrt{2}-1}{\sqrt{2}+1} \times \dfrac{\sqrt{2}-1}{\sqrt{2}-1} = \dfrac{(\sqrt{2}-1)(\sqrt{2}-1)}{(\sqrt{2}+1)(\sqrt{2}-1)} = \dfrac{2+1-2\sqrt{2}}{2-1} = 3 - 2\sqrt{2}$。

答案 A

练 习

1.

Quantity A	Quantity B
$\sqrt{r^6}$	r^3

A. Quantity A is greater.

B. Quantity B is greater.

C. The two quantities are equal.

D. The relationship cannot be determined from the information given.

2. $a > b > c > 0$

Quantity A	Quantity B
$\sqrt{ab}\sqrt{c}$	$\sqrt{a}\sqrt{bc}$

A. Quantity A is greater.

B. Quantity B is greater.

C. The two quantities are equal.

D. The relationship cannot be determined from the information given.

3. Which of the following double inequalities involving $\sqrt[3]{0.8}$ is true?

   A. $0 < \sqrt[3]{0.8} < 0.2$
   B. $0.2 < \sqrt[3]{0.8} < 0.4$
   C. $0.4 < \sqrt[3]{0.8} < 0.6$
   D. $0.6 < \sqrt[3]{0.8} < 0.8$
   E. $0.8 < \sqrt[3]{0.8} < 1.0$

4. If $x > 0$ and $\dfrac{10}{14} = \sqrt{\dfrac{25}{x}}$, what is the value of x?

5. What is the nearest integer to the value of $\sqrt[3]{27+64}$?

答案及解析

1.

Quantity A	Quantity B
$\sqrt{r^6}$	r^3

   A. Quantity A is greater.
   B. Quantity B is greater.
   C. The two quantities are equal.
   D. The relationship cannot be determined from the information given.

解题思路

$\sqrt{r^6} = |r^3|$,

当 $r \geq 0$ 时, $\sqrt{r^6} = r^3$,

当 $r < 0$ 时, $\sqrt{r^6} > 0 > r^3$。

由于题干没有给出 r 与 0 的大小关系，因此 $\sqrt{r^6}$ 与 $|r^3|$ 大小无法确定。

答案 D

2. $a > b > c > 0$

Quantity A	Quantity B
$\sqrt{ab}\sqrt{c}$	$\sqrt{a}\sqrt{bc}$

   A. Quantity A is greater.
   B. Quantity B is greater.
   C. The two quantities are equal.
   D. The relationship cannot be determined from the information given.

解题思路

$\sqrt{ab}\sqrt{c} = \sqrt{a}\sqrt{bc} = \sqrt{abc}$，

数量 A 与数量 B 相等。

答案 C

3. Which of the following double inequalities involving $\sqrt[3]{0.8}$ is true?

   A. $0 < \sqrt[3]{0.8} < 0.2$
   B. $0.2 < \sqrt[3]{0.8} < 0.4$
   C. $0.4 < \sqrt[3]{0.8} < 0.6$
   D. $0.6 < \sqrt[3]{0.8} < 0.8$
   E. $0.8 < \sqrt[3]{0.8} < 1.0$

解题思路

如果 $0 < a < 1$，a^n 的值随着 n 变大而变小，

因此，$0.8^1 < \sqrt[3]{0.8}(=0.8^{\frac{1}{3}}) < 0.8^0(=1)$。

答案 E

4. If $x > 0$ and $\dfrac{10}{14} = \sqrt{\dfrac{25}{x}}$, what is the value of x?

解题思路

$\dfrac{10}{14} = \dfrac{5}{7} = \sqrt{\dfrac{25}{x}} = \dfrac{5}{\sqrt{x}}$，

则 $\sqrt{x} = 7$，$x = 49$。

答案 49

5. What is the nearest integer to the value of $\sqrt[3]{27+64}$?

解题思路

因为 $27 + 64 = 91$，

所以本题是要找与 91 最接近的立方数。

$4^3 = 64$，与 91 的距离 $= |64 - 91| = 27$，

$5^3 = 125$，与 91 的距离 $= |125 - 91| = 34$，

因此，与 91 最接近的立方数为 4^3，

与 $\sqrt[3]{27+64}$ 最接近的整数为 4。

答案 4

2.4 解线性方程

基本词汇

solve an solution 解方程   equivalent equations 同解方程   linear equation 线性方程

概念

解方程就是去寻找能够让等式成立的未知数的值。

两个方程有相同的解则被称为同解方程。例如：$x+2=1$ 和 $2x+4=2$ 是同解方程，它们的解都是 $x=-1$。

我们在解方程时，要用到两个方程的同解原理（等式的基本性质）。

（1）方程两边同时加上或减去相同的常数，方程的解不变。

（2）方程两边同时乘以或除以一个非零数字，方程的解不变。

线性方程也称一次方程式，指的是未知数都是一次的方程。

一元一次方程：$ax+b=c$。

二元一次方程：$ax+by=c$。

一个方程中含有2个未知数，通常无法求解。有几个未知数，一般需要联立几个方程才能求解。

解含有一个未知数的线性方程，我们通常会使用合并同类项和移项的方法来求解。

解含有两个未知数的线性方程，我们通常会用换元法和消元法。

$3x + y = 4$
$x - 2y = -1$

Quantity A	Quantity B
x	y

A. Quantity A is greater.
B. Quantity B is greater.
C. The two quantities are equal.

D. The relationship cannot be determined from the information given.

解题思路

本题让我们比较 x 与 y 的大小。

方程有 2 个未知数，因此联立 2 个不同的方程来求解。

联立方程组：
$$\begin{cases} 3x + y = 4 \\ x - 2y = -1 \end{cases}$$

解得：$x = 1$，$y = 1$。

因此，数量 A = 数量 B。

答案　C

$3x + 2y = 8$

x and y are integers and $x > y$.

Quantity A　　　　Quantity B

$3x + 4y$　　　　　8

A. Quantity A is greater.

B. Quantity B is greater.

C. The two quantities are equal.

D. The relationship cannot be determined from the information given.

题干翻译

$3x + 2y = 8$，

x 和 y 是整数，且 $x > y$。

解题思路

本题要我们比较 $3x + 4y$ 与 8 的大小关系。

一个方程两个未知数，通常无法求解，因此我们考虑用试数的方法。

注意，题干只给出 x 和 y 是整数，且 $x > y$，没有说明 y 的正负。

如果 $x = 2$，$y = 1$，则 $3x + 4y = 10 > 8$，

如果 $x = 4$，$y = -2$，则 $3x + 4y = 4 < 8$，

因此，$3x + 4y$ 与 8 的大小关系无法确定。

答案　D

练习

1. $x = 6 + c$

$c = y + 6$

Quantity A	Quantity B
$x + y$	$2c$

A. Quantity A is greater.

B. Quantity B is greater.

C. The two quantities are equal.

D. The relationship cannot be determined from the information given.

2. $(x+3)(y-4) = 0$

Quantity A	Quantity B
xy	-12

A. Quantity A is greater.

B. Quantity B is greater.

C. The two quantities are equal.

D. The relationship cannot be determined from the information given.

答案及解析

1. $x = 6 + c$

$c = y + 6$

Quantity A	Quantity B
$x + y$	$2c$

A. Quantity A is greater.

B. Quantity B is greater.

C. The two quantities are equal.

D. The relationship cannot be determined from the information given.

解题思路

本题让我们比较 $x+y$ 与 $2c$ 的大小。

根据 $c = y + 6$ 可知，$y = c - 6$，

$x + y = 6 + c + c - 6 = 2c$，

故数量 A 与数量 B 相等。

答案 C

2. $(x+3)(y-4) = 0$

Quantity A	Quantity B
xy	-12

A. Quantity A is greater.

B. Quantity B is greater.

C. The two quantities are equal.

D. The relationship cannot be determined from the information given.

解题思路

本题要求我们比较 xy 与 -12 的大小关系。

$(x+3)(y-4) = 0$，

所以，$x = -3$ 或 $y = 4$，但并没有要求两者同时成立。

如果 $x = -3$，$y = 4$，$xy = -12$，

如果 $x = -3$，$y = -4$，$xy = 12 > -12$，

因此 xy 与 -12 的大小关系无法确定。

答案 D

2.5 解一元二次方程

基本词汇

quadratic equation 二次方程
quadratic formula 求根公式
solution 解
root 根

概念

通过化简后，只含有一个未知数（一元），并且未知数的最高次数是 2（二次）的整式方程，叫作一元二次方程。

含有未知数 x 的一元二次方程可以写成以下形式：

$ax^2 + bx + c = 0$（a，b，c 都是实数，且 $a \neq 0$）

考点1 一元二次方程 $ax^2+bx+c=0$ 求解

当一元二次方程有实数根时，实数根可以用求根公式得出：

$$x = \frac{-b \pm \sqrt{b^2 - 4ac}}{2a}。$$

除了求根公式，我们还可以用因式分解法来求解。因式分解一般会用到十字相乘法。

如果一元二次方程 $ax^2+bx+c=0$ 存在两个实根 x_1，x_2，那么它可以因式分解为 $a(x-x_1)(x-x_2)=0$。

考点2 一元二次方程 $ax^2+bx+c=0$ 的性质

（1）一元二次方程有几个解？

我们一般用 $\triangle = b^2 - 4ac$ 来判断一元二次方程的根的数量。

- 当 $\triangle = b^2 - 4ac > 0$，一元二次方程有两个不同的实数根。
- 当 $\triangle = b^2 - 4ac = 0$，一元二次方程有一个实数根。
- 当 $\triangle = b^2 - 4ac < 0$，一元二次方程没有实数根。

（2）两根之和 $x_1 + x_2 = -\dfrac{b}{a}$，两根之积 $x_1 x_2 = \dfrac{c}{a}$。

例题 01.

Quantity A	Quantity B
The number of integer solutions of the equation $3x^2+5x-2=0$	1

A. Quantity A is greater.
B. Quantity B is greater.
C. The two quantities are equal.
D. The relationship cannot be determined from the information given.

解题思路

注意，本题要求我们比较的是方程 $3x^2+5x-2=0$ 的"整数解"的个数和 1 的大小关系，所以需要老老实实算出 $3x^2+5x-2=0$ 的解，而不是用 $\triangle=b^2-4ac$ 来判断。

$3x^2+5x-2=0$，
$(3x-1)(x+2)=0$，
$x_1=\dfrac{1}{3}$，$x_2=-2$。

虽然这个方程有 2 个解，但这个方程的整数解只有 -2 一种情况，所以数量 A 为 1，与数量 B 相等。

答案 C

例题 02.

If the equation $x^2-(m-2)x-(n-4)^2=0$ has only one solution, what is the product of m and n?

A. 4　　B. 8　　C. 10　　D. 12　　E. 16

题干翻译

如果方程 $x^2-(m-2)x-(n-4)^2=0$ 只有一个解，那么 m 和 n 的乘积是多少？

解题思路

一元二次方程只有 1 个解，说明 $\triangle=b^2-4ac=0$。

对于 $x^2-(m-2)x-(n-4)^2=0$ 来说，$a=1$，$b=-(m-2)$，$c=-(n-4)^2$（对应形式：$ax^2+bx+c=0$），

$\triangle=b^2-4ac=(m-2)^2+4(n-4)^2=0$。

因为 $(m-2)^2$ 与 $(n-4)^2$ 都是偶数次方，而偶数次方 ≥ 0，

因此 $m-2=0$，且 $n-4=0$，

$m=2$，$n=4$，

$mn=8$。

答案 B

例题 03.

If the function $f(x) = x^2 + px + 9$ has the roots: x_1 and x_2, $x_1 + x_2 = 6$. What is the value of p^2?

题干翻译

如果函数 $f(x) = x^2 + px + 9$ 有 x_1 和 x_2 两个根，且 $x_1 + x_2 = 6$，p^2 的值是多少？

解题思路

对于函数 $f(x) = x^2 + px + 9$ 来说（对应形式：$ax^2 + bx + c = 0$），

$a = 1$，$b = p$，$c = 9$，

$x_1 + x_2 = -\dfrac{b}{a} = -p = 6$，

$p = 6$，

因此 $p^2 = 36$。

答案 36

练习

1. a and b are the two solutions of the equation $x^2 = x + 2$.

Quantity A	Quantity B
$a + b$	0

   A. Quantity A is greater.

   B. Quantity B is greater.

   C. The two quantities are equal.

   D. The relationship cannot be determined from the information given.

2. If $x = 2$ is one solution of the equation $2x^2 - 3x - k = 0$, where k is a constant, what is the other solution of the equation?

   A. $x = -\dfrac{1}{4}$ B. $x = -\dfrac{1}{2}$ C. $x = \dfrac{1}{4}$ D. $x = \dfrac{1}{2}$ E. $x = \dfrac{2}{3}$

答案及解析

1. *a* and *b* are the two solutions of the equation $x^2 = x + 2$.

   Quantity A   Quantity B

     $a + b$ 0

   A. Quantity A is greater.

   B. Quantity B is greater.

   C. The two quantities are equal.

   D. The relationship cannot be determined from the information given.

题干翻译

a 和 b 是方程 $x^2 = x + 2$ 的两个解。

解题思路

本题需要我们比较 $a + b$ 与 0 的大小关系。

$x^2 = x + 2$，

$x^2 - x - 2 = 0$，

$(x - 2)(x + 1) = 0$，

$x = 2$ 或 $x = -1$。

两个解为 2 和 -1，$a + b = 2 - 1 = 1 > 0$，所以数量 A 更大。

答案   A

2. If $x = 2$ is one solution of the equation $2x^2 - 3x - k = 0$, where k is a constant, what is the other solution of the equation?

   A. $x = -\dfrac{1}{4}$ B. $x = -\dfrac{1}{2}$ C. $x = \dfrac{1}{4}$ D. $x = \dfrac{1}{2}$ E. $x = \dfrac{2}{3}$

题干翻译

如果 $x = 2$ 是方程 $2x^2 - 3x - k = 0$ 的一个解，其中 k 是常数，那么方程的另一个解是什么？

解题思路

对于方程 $2x^2 - 3x - k = 0$ 来说，$a = 2$，$b = -3$，$c = -k$，

两根之和 $x_1 + x_2 = -\dfrac{b}{a} = \dfrac{3}{2}$。

根据题干条件可知其中一个根为 2，代入上述等式可得：$2 + x_2 = \dfrac{3}{2}$，

$x_2 = -\dfrac{1}{2}$。

答案   B

2.6 解不等式

基本词汇

inequality 不等式

概念

使用下列不等式符号的数学表达式称为不等式。

小于号 <（less than），

大于号 >（greater than），

小于等于号 ≤（less than or equal to），

大于等于号 ≥（greater than or equal to）。

解不等式的方法与解方程类似，主要通过合并同类项和移项来简化不等式。

不等式的基本性质有：

① 不等式两侧同时加上或减去同一个数之后，不等号方向不变。

② 不等式两侧同时乘以或除以同一个正数之后，不等号方向不变。

③ 不等式两侧同时乘以或除以同一个负数之后，不等号方向改变。

考点1 解线性不等式

不等式常用公式：

若 $a > b$，$b > c$，则 $a > c$，

若 $a > b$，则 $a + c > b + c$，

若 $a > b$，$c > d$，则 $a + c > b + d$，

若 $a > b$，$c > 0$，则 $ac > bc$，

若 $a > b$，$c < 0$，则 $ac < bc$，

若 $a > b > 0$，$c > d > 0$，则 $ac > bd$，

若 $a > b > 0$，且 n 为大于 1 的整数，则 $a^n > b^n$，

若 $a > b > 0$，且 n 为大于 1 的整数，则 $\sqrt[n]{a} > \sqrt[n]{b}$。

例题 01.

If $r > t$, which of the following inequalities must be satisfied?
Indicate all such inequalities.

A. $r - t > 0$ B. $t - r > 0$ C. $t - r < 0$ D. $r - t < 0$ E. $r + t > 0$

题干翻译

如果 $r > t$，下列哪个不等式一定成立？

解题思路

根据题干 $r > t$ 条件可知：

$r - t > 0$（不等式两边同时减去 t，不等号方向不变），A 选项正确。

根据 $r - t > 0$ 可知，$t - r < 0$（不等式两边同时乘以 -1，不等号方向改变），C 选项正确。

答案 AC

例题 02.

$a < 0$，$\dfrac{b}{a} > 1$

Quantity A	Quantity B
$a - b$	0

A. Quantity A is greater.

B. Quantity B is greater.

C. The two quantities are equal.

D. The relationship cannot be determined from the information given.

解题思路

根据题干条件 $a < 0$，$\dfrac{b}{a} > 1$ 可知，$b < a$（不等式两边同时乘以一个相同的负数，不等号方向改变），

因此 $a - b > 0$，

数量 A 更大。

答案 A

考点 2 解一元高次不等式

解一元高次不等式，我们通常会用穿针引线法。

穿针引线法步骤如下：

第一步：先通过移项，使不等式右侧为 0，然后对左侧进行分解因式。

第二步：将不等号换成等号，解出所有的根。

第三步：在数轴上从左到右依次标出各根。

第四步：画穿根线：从最大值的右上方开始，向左开始画线，经过一个根，就穿一次数轴，依次穿过各根。

如果不等号为">"，则取数轴上方穿根线所在的范围；

如果不等号为"<"，则取数轴下方穿根线所在的范围。

式子越复杂，穿针引线法越好用。

例题

$x^2 - 2x < 3$

解题思路

（1）移项，使不等式右侧为 0：$x^2 - 2x - 3 < 0$。

（2）将不等号换成等号，解出所有的根。$x^2 - 2x - 3 = (x-3)(x+1) = 0$，两个根为 3 和 -1。

（3）在数轴上从左到右依次标出各根。

（4）从右上方开始穿针引线，经过一个解穿一次轴。

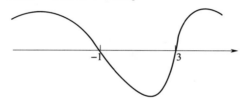

因为不等式的符号为 <0，所以看数轴下方穿根线所在的区域，即 $-1 < x < 3$。

注意

① 保证分解因式后，x 前的系数为正数或者保证 x 最高次方项前的系数为正数。

② 奇穿偶不穿：因式分解后带偶数次方的因式解得的根，不能穿线。

考点3　解分式不等式

分式不等式，我们依旧可以用穿针引线法来解，步骤如下：

第一步：先通过移项，使不等式右侧为 0。

第二步：对分式不等式进行通分，通分之后分子乘以分母。

第三步：因式分解，将不等号换成等号，解出所有的根。

第四步：在数轴上从左到右依次标出各根，接着画穿根线：从最大值的右上方开始，向左开始画线，经过一个根，就穿一次数轴，依次穿过各根。

如果不等号为">"，则取数轴上方穿根线所在的范围；

如果不等号为"<"，则取数轴下方穿根线所在的范围。

注意，因为数学中规定，分数分母不能为 0，所以如果不等式涉及等号，需要把分母为 0 的情况去除。

> **例题**

$\dfrac{2x^2+3x-7}{x^2-x-2} \geq 1$。

解题思路

(1) 通过移项，使不等式右侧为0：$\dfrac{2x^2+3x-7}{x^2-x-2} - 1 \geq 0$。

(2) 通分：$\dfrac{2x^2+3x-7-(x^2-x-2)}{x^2-x-2} \geq 0$，即 $\dfrac{x^2+4x-5}{x^2-x-2} \geq 0$。

分子乘以分母：$(x^2+4x-5)(x^2-x-2) \geq 0$（乘除不改变正负）。

(3) 因式分解，求根。$(x-1)(x+5)(x-2)(x+1) \geq 0$

(4) 穿针引线：

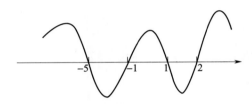

由于该不等式的符号是 ≥ 0，所以看数轴线上方的解，同时要去除分母为0的解 $x=2$ 和 $x=-1$。

因此这个不等式的结果为：$x \leq -5$，$-1 < x \leq 1$，$x > 2$。

考点4　解绝对值不等式

解法：分类讨论。

第一步：分两种情况来分类讨论，去掉绝对值符号。

第二步：对两种情况分别进行求解，最后结果取并集。

> **例题**

$3 < |x-5| \leq 8$

解题思路

可能性1：$x-5 \geq 0$，即 $x \geq 5$ 时，绝对值符号可直接去掉。

即 $3 < x-5 \leq 8$，解得 $8 < x \leq 13$，由于这个解的范围均满足 $x \geq 5$，所以 $8 < x \leq 13$。

可能性2：$x-5 < 0$，即 $x < 5$，去绝对值符号时，绝对值下的数值要乘以 -1。

即 $3 < -(x-5) \leq 8$，

各个式子同时乘以 -1，即 $-3 > x-5 \geq -8$，

解得 $2 > x \geq -3$，由于这个解的范围均满足 $x < 5$，所以 $-3 \leq x < 2$。

对可能性1和可能性2的结果取并集，得出 x 的取值范围为 $-3 \leq x < 2$，$8 < x \leq 13$。

练习

1. Which of following inequalities is equivalent to $3 \leqslant x \leqslant 4$?

    A. $x^2 \leqslant 7x - 12$ B. $x^2 \leqslant 7x + 12$ C. $x^2 \leqslant 12 - 7x$ D. $x^2 \geqslant 7x - 12$ E. $x^2 \geqslant 12 - 7x$

2. For which of the following values of x is the inequality $\dfrac{(x-4)^4 (x+5)^6}{(x-2)^{10}} \leqslant 0$ satisfied? Indicate <u>all</u> such values.

    A. -9 B. -5 C. -2 D. 0 E. 1

    F. 4 G. 6

3.

The solution set of which of the following inequalities is graphed on the number line shown?

    A. $|x| \leqslant 12$ B. $2 \leqslant |x| \leqslant 12$ C. $|x - 2| \leqslant 10$ D. $|x + 7| \leqslant 5$ E. $|x - 7| \leqslant 5$

答案及解析

1. Which of following inequalities is equivalent to $3 \leqslant x \leqslant 4$?

    A. $x^2 \leqslant 7x - 12$ B. $x^2 \leqslant 7x + 12$ C. $x^2 \leqslant 12 - 7x$ D. $x^2 \geqslant 7x - 12$ E. $x^2 \geqslant 12 - 7x$

题干翻译

下列哪个不等式的解为 $3 \leqslant x \leqslant 4$?

解题思路

A 选项：$x^2 \leqslant 7x - 12$,

           $x^2 - 7x + 12 \leqslant 0$,

           $(x-3)(x-4) \leqslant 0$,

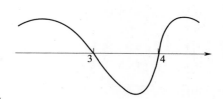

由于该不等式的符号是 $\leqslant 0$，所以取数轴线下方的解 $3 \leqslant x \leqslant 4$。

该解符合题干要求，故 A 正确。

答案   A

2. For which of the following values of x is the inequality $\dfrac{(x-4)^4(x+5)^6}{(x-2)^{10}} \leq 0$ satisfied?

Indicate all such values.

A. -9 B. -5 C. -2 D. 0
E. 1 F. 4 G. 6

题干翻译

下列哪个 x 的值符合不等式 $\dfrac{(x-4)^4(x+5)^6}{(x-2)^{10}} \leq 0$?

解题思路

由于每个表达式都是偶数次方，因此都 ≥ 0，

且分数式分母不能为 0，

因此只能分子为 0，

当 $x = 4$ 或 -5 的时候，分子为 0。

答案 B F

3.

The solution set of which of the following inequalities is graphed on the number line shown?

A. $|x| \leq 12$ B. $2 \leq |x| \leq 12$ C. $|x-2| \leq 10$ D. $|x+7| \leq 5$ E. $|x-7| \leq 5$

题干翻译

下列哪个不等式的解如数轴所示？

解题思路

数轴上的解的范围为：$-12 \leq x \leq -2$，

D 选项：$|x+7| \leq 5$，

可能性 1：$x+7 \geq 0$，即 $x \geq -7$，绝对值符号直接去除，得到 $x+7 \leq 5$，解得 $x \leq -2$，

由于 $x \leq -2$ 满足 $x \geq -7$，因此 $x \leq -2$。

可能性 2：$x+7 < 0$，即 $x < -7$，去绝对值符号时，需要乘以 -1，得到 $-(x+7) \leq 5$，

再同时乘以 -1，得到 $x+7 \geq -5$，解得 $x \geq -12$，

由于不等式需要满足 $x < -7$，因此 $-12 \leq x < 7$。

取可能性 1 与可能性 2 的并集，x 的解的范围为：$-12 \leq x \leq -2$。

答案 D

2.7 数列

基本词汇

arithmetic sequence 等差数列 set 几何 term 项
geometric sequence 等比数列 subset 子集

考点 1 等差数列

等差数列的概念：从第二项起，每一项与它前一项的差等于同一个常数的一种数列。这个常数叫作等差数列的公差，常用字母 d 表示。

等差数列的通项公式：$a_n = a_1 + (n-1)d$。

等差数列的求和公式：$S_n = \dfrac{(a_1 + a_n)n}{2}$。

考点 2 等比数列

等比数列的概念：从第二项起，每一项与它的前一项的比值等于同一个常数的一种数列。这个常数叫作等比数列的公比，通常用字母 q 表示（$q \neq 0$）。

等比数列的通项公式：$a_n = a_1 q^{n-1}$。

等差数列的求和公式：$S_n = \dfrac{a_1(1-q^n)}{1-q}$。

考点 3 非等差非等比数列

在 GRE 数学考试中，非等差非等比数列的题目会直接给公式，我们根据题目中给出的公式进行相应的计算即可。

例题 01.

The first term of an infinite sequence is 4, and each term after the first term is 7 greater than the preceding term. What is the 64th term of the sequence?

A. 188　　　　B. 431　　　　C. 445　　　　D. 448　　　　E. 452

题干翻译

无穷数列的第一项是 4，第一项之后的每一项都比前一项大 7。数列的第 64 项是什么？

解题思路

根据题干条件可知，这是一个首项 a_1 为 4、公差 d 为 7 的等差数列。

因此，第 64 项 $a_{64} = a_1 + (64-1)d = 4 + 63 \times 7 = 445$。

答案 C

a_1，a_2，a_3，$\cdots$，a_{99}

In the sequence shown, each term after the first is 1 greater than the preceding term. The sum of the 99 terms of the sequence is 99. What is the value of the first term of the sequence?

A. -52　　　B. -51　　　C. -50　　　D. -49　　　E. -48

题干翻译

在所示的数列 a_1，a_2，a_3，$\cdots$，a_{99} 中，第一项之后的每一项都比前一项大 1。数列的 99 项之和是 99。数列第一项的值是多少？

解题思路

根据题干可知，这是一个公差 d 为 1 的等差数列。

$a_{99} = a_1 + (99-1)d = a_1 + 98$，

前 99 项的和 $S_{99} = \dfrac{(a_1 + a_{99}) \times 99}{2} = 99$，

$\dfrac{(a_1 + a_1 + 98) \times 99}{2} = 99$，

$2a_1 + 98 = 2$，

$2a_1 = -96$，

$a_1 = -48$。

答案 E

x_1，x_2，x_3，$\cdots$，x_j，$\cdots$

The sequence shown is defined by $x_1 = 2$, and $x_{j+1} = \dfrac{1}{2}x_j$ for each positive integer j.

Quantity A　　　　　Quantity B

x_9　　　　　　　$(2^{13})x_{22}$

A. Quantity A is greater.
B. Quantity B is greater.

C. The two quantities are equal.

D. The relationship cannot be determined from the information given.

题干翻译

对于数列 x_1，x_2，x_3，$\cdots$，x_j，$\cdots$，$x_1=2$，且对每一个正整数 j 来说 $x_{j+1}=\dfrac{1}{2}x_j$。

解题思路

本题让我们比较 x_9 与 $(2^{13})x_{22}$ 的大小关系。

根据题干可知，该数列是一个首项为 2、公比为 $\dfrac{1}{2}$ 的等比数列。

$x_9 = x_1 q^8 = 2 \times \left(\dfrac{1}{2}\right)^8 = \left(\dfrac{1}{2}\right)^7$，

$(2^{13})x_{22} = (2^{13})x_1 q^{21} = (2^{13}) \times 2 \times \left(\dfrac{1}{2}\right)^{21} = \left(\dfrac{1}{2}\right)^7$，

数量 A = 数量 B。

答案　C

A sequence of numbers P_1, P_2, P_3, $\cdots$, P_n, $\cdots$ is defined as follows: $P_1=1$, $P_2=2$, and $P_n=4\dfrac{P_{n-1}}{P_{n-2}}$ for each integer n greater than 2. What is the value of P_4?

A. 4　　　　B. 6　　　　C. 12　　　　D. 16　　　　E. 24

题干翻译

数列 P_1，P_2，P_3，$\cdots$，P_n，$\cdots$ 定义如下：对于大于 2 的每个整数 n，$P_1=1$，$P_2=2$，$P_n=4\dfrac{P_{n-1}}{P_{n-2}}$。$P_4$ 的值是多少？

解题思路

根据题干条件可知，该数列是一个非等差非等比数列，具体的项根据题干要求代入即可。

$P_4 = 4\dfrac{P_3}{P_2}$，

$P_3 = 4\dfrac{P_2}{P_1} = 8$，

将 P_3 的数值代入 $P_4 = 4\dfrac{P_3}{P_2} = 4 \times \dfrac{8}{2} = 16$。

答案　D

练 习

1. 10, 13, 16, 19, 22, ⋯

The first term in the sequence shown is 10, and each term after the first is 3 greater than the previous term. What is the value of the 92nd term in the sequence?

A. 273   B. 276   C. 280   D. 283   E. 286

2. An office building has 6 floors. If there are n offices on the top floor and each floor has 3 more offices than the floor just above it, how many offices are there in the building?

A. $n - 18$ B. $n - 45$ C. $6n + 18$ D. $6n - 45$ E. $6n + 45$

3. $T_1, T_2, T_3, \cdots, T_k, \cdots$

The sequence shown is defined by $T_1 = 2$ and $T_{k+1} = \frac{1}{3} T_k$ for each positive integer k.

Quantity A	Quantity B
T_5	$(3^{11}) T_{16}$

A. Quantity A is greater.

B. Quantity B is greater.

C. The two quantities are equal.

D. The relationship cannot be determined from the information given.

4. $a_k = \left(\frac{1}{k} - \frac{1}{k+1} \right)$ for any positive integer k

Quantity A	Quantity B
$a_3 + a_4 + a_5 + a_6 + a_7$	$\frac{1}{8}$

A. Quantity A is greater.

B. Quantity B is greater.

C. The two quantities are equal.

D. The relationship cannot be determined from the information given.

答案及解析

1. 10, 13, 16, 19, 22, …

The first term in the sequence shown is 10, and each term after the first is 3 greater than the previous term. What is the value of the 92nd term in the sequence?

A. 273　　　　B. 276　　　　C. 280　　　　D. 283　　　　E. 286

题干翻译

10, 13, 16, 19, 22, …

上述数列的第一项是10，第一项之后的每一项都比前一项大3。序列中第92项的值是多少？

解题思路

根据题干可知，这个数列是一个首项 a_1 为10、公差 d 为3的等差数列。

$a_{92} = a_1 + (92-1)d = 10 + 91 \times 3 = 283$。

答案 D

2. An office building has 6 floors. If there are n offices on the top floor and each floor has 3 more offices than the floor just above it, how many offices are there in the building?

A. $n-18$　　　B. $n-45$　　　C. $6n+18$　　　D. $6n-45$　　　E. $6n+45$

题干翻译

一栋办公楼有6层。如果顶层有 n 间办公室，每层比上面的楼层多3间办公室，那么大楼里有多少间办公室？

解题思路

根据题干可知，这是一个首项为 $a_1 = n$、公差 d 为3的等差数列。

$a_6 = a_1 + (6-1)d = n + 15$，

因此，这6层办公室的总间数 $S_6 = \dfrac{(a_1 + a_6) \times 6}{2} = \dfrac{(n + n + 15) \times 6}{2} = 6n + 45$。

答案 E

3. $T_1, T_2, T_3, \cdots, T_k, \cdots$

The sequence shown is defined by $T_1 = 2$ and $T_{k+1} = \dfrac{1}{3} T_k$ for each positive integer k.

Quantity A	Quantity B
T_5	$(3^{11})T_{16}$

A. Quantity A is greater.

B. Quantity B is greater.

C. The two quantities are equal.

D. The relationship cannot be determined from the information given.

题干翻译

对于数列 T_1，T_2，T_3，$\cdots$，T_k，$\cdots$，$T_1 = 2$，且对每一个正整数 k 来说，$T_{k+1} = \dfrac{1}{3}T_k$。

解题思路

本题让我们比较的是 T_5 和 $(3^{11})T_{16}$ 的大小关系。

根据题干信息可知，这是一个首项 $T_1 = 2$、公比为 $\dfrac{1}{3}$ 的等比数列。

因此，$T_5 = T_1 q^4 = 2 \times \left(\dfrac{1}{3}\right)^4$，

$(3^{11})T_{16} = (3^{11})T_1 q^{15} = (3^{11}) \times 2 \times \left(\dfrac{1}{3}\right)^{15} = 2 \times \left(\dfrac{1}{3}\right)^4$，

即数量 A = 数量 B。

答案 C

4. $a_k = \left(\dfrac{1}{k} - \dfrac{1}{k+1}\right)$ for any positive integer k

Quantity A	Quantity B
$a_3 + a_4 + a_5 + a_6 + a_7$	$\dfrac{1}{8}$

A. Quantity A is greater.

B. Quantity B is greater.

C. The two quantities are equal.

D. The relationship cannot be determined from the information given.

题干翻译

对每一个正整数 k 来说，$a_k = \dfrac{1}{k} - \dfrac{1}{k+1}$。

解题思路

本题需要我们比较 $a_3 + a_4 + a_5 + a_6 + a_7$ 与 $\dfrac{1}{8}$ 的大小关系。

根据题干通项公式可知:

$a_3 = \dfrac{1}{3} - \dfrac{1}{4}$,

$a_4 = \dfrac{1}{4} - \dfrac{1}{5}$,

$a_5 = \dfrac{1}{5} - \dfrac{1}{6}$,

$a_6 = \dfrac{1}{6} - \dfrac{1}{7}$,

$a_7 = \dfrac{1}{7} - \dfrac{1}{8}$,

$a_3 + a_4 + a_5 + a_6 + a_7 = \dfrac{1}{3} - \dfrac{1}{4} + \dfrac{1}{4} - \dfrac{1}{5} + \dfrac{1}{5} - \dfrac{1}{6} + \dfrac{1}{6} - \dfrac{1}{7} + \dfrac{1}{7} - \dfrac{1}{8}$,

$= \dfrac{1}{3} - \dfrac{1}{8}$,

$= \dfrac{5}{24} > \dfrac{3}{24}\left(=\dfrac{1}{8}\right)$。

因此数量 A 更大。

答案 A

2.8 函数

基本词汇

function 函数　　　　　　　domain 定义域

概念

有一个未知数的代数式可以被用来定义那个未知数的函数。函数通常用字母如 f, g, h 来表示。例如，代数式 $2x+5$ 可以被用来定义一个函数 $f(x)=2x+5$。在这个函数中，$f(x)$ 被称为 f 在 x 时的值，这个值可以通过把 x 的值代入函数表达式中得出。例如，如果 $x=2$，$f(2)=9$。

函数题除了考查根据函数表达式，求变量 x 等于某个常数时，函数 $f(x)$ 对应的数值，也有可能考查变量的相关替换。例如，如果 $f(x)=2x-1$，那么 $f(3x+1)$ 表示的就是把原式中的所有未知数 x 替换成 $3x+1$，所以 $f(3x+1)=2(3x+1)-1=6x+1$。

在函数中，所有可能的自变量的值的集合被称为函数的定义域。

注意，$f(x)$ 和 $g(x)$ 仅仅是函数常用的表达符号，在 GRE 数学考试中，ETS 考查的函数题中会有大量的新定义的运算符号构成的函数。例如，定义 $a \odot b = ab^2$，那么 $2 \odot 3 = 2 \times 3^2 = 18$。

例题

$x^{..}$ is defined as the 3-digit integer formed by reversing the digits of integer x; for instance, $258^{..}$ is equal to 852. R is a 3-digit integer such that its units digit is 2 greater than its hundreds digit.

Quantity A	Quantity B
$R^{..} - R$	200

A. Quantity A is greater.
B. Quantity B is greater.
C. The two quantities are equal.
D. The relationship cannot be determined from the information given.

题干翻译

$x^{..}$ 被定义为通过反转整数 x 数位上的数字形成的 3 位整数；比如，$258^{..}$ 等于 852。R 是 3 位数整数，其个位对应的数值比百位对应的数值大 2。

解题思路

本题要求比较 $R^{..} - R$ 与 200 的大小关系。

设三位数 R 为 abc（a 为百位，b 为十位，c 为个位），$R = 100a + 10b + c$，

且根据题干条件"R 个位对应的数值比百位对应的数值大 2"可知，$c = a + 2$。

则根据题干定义可知，$R^{..} = cba = 100c + 10b + a$，

$R^{..} - R = 100c + 10b + a - (100a + 10b + c)$。

将 $c = a + 2$ 代入上面的表达式可得：

$R^{..} - R = 100(a+2) + 10b + a - (100a + 10b + a + 2) = 200 - 2 = 198 < 200$，

即数量 B 更大。

答案 B

练 习

1. The operation ∅ is defined for all numbers c and d by $c \emptyset d = c^2 + d^2$.

Quantity A	Quantity B
$(3 \emptyset 5) \emptyset 7$	$3 \emptyset (5 \emptyset 7)$

A. Quantity A is greater.

B. Quantity B is greater.

C. The two quantities are equal.

D. The relationship cannot be determined from the information given.

2. The function f is defined by $f(n) = \dfrac{2n-1}{2n+1}$ for all positive integers n. What is the least positive integers m for which the product $(f(1))(f(2))\cdots(f(m))$ is less than or equal to $\dfrac{1}{15}$?

A. 6   B. 7   C. 8   D. 14   E. 15

3. The function f is defined for all numbers x by $f(2x) = x^2 - 2x + 8$.

Quantity A	Quantity B
$f(6)$	12

A. Quantity A is greater.

B. Quantity B is greater.

C. The two quantities are equal.

D. The relationship cannot be determined from the information given.

答案及解析

1. The operation Ø is defined for all numbers c and d by $c \text{ Ø } d = c^2 + d^2$.

Quantity A	Quantity B
(3 Ø 5) Ø 7	3 Ø (5 Ø 7)

A. Quantity A is greater.

B. Quantity B is greater.

C. The two quantities are equal.

D. The relationship cannot be determined from the information given.

题干翻译

对所有数字 c 和 d，有运算符号 Ø，使得 $c \text{ Ø } d = c^2 + d^2$。

解题思路

本题需要我们比较的是 (3 Ø 5) Ø 7 与 3 Ø (5 Ø 7) 的大小关系。

(3 Ø 5) Ø 7 = $(3^2 + 5^2)$ Ø 7 = 34 Ø 7 = $34^2 + 7^2$ = 1205，

3 Ø (5 Ø 7) = 3 Ø $(5^2 + 7^2)$ = 3 Ø 74 = $3^2 + 74^2$ = 5485 > 1205，

即数量 B 更大。

答案 B

2. The function f is defined by $f(n) = \dfrac{2n-1}{2n+1}$ for all positive integers n. What is the least positive integers m for which the product $(f(1))(f(2))\cdots(f(m))$ is less than or equal to $\dfrac{1}{15}$?

A. 6 B. 7 C. 8 D. 14 E. 15

题干翻译

对于所有正整数 n，有函数 $f(n) = \dfrac{2n-1}{2n+1}$。使得 $(f(1))(f(2))\cdots(f(m)) \leq \dfrac{1}{15}$ 的最小正整数 m 是多少？

解题思路

$(f(1))(f(2))\cdots(f(m)) \leq \dfrac{1}{15}$,

$\dfrac{1}{3} \times \dfrac{3}{5} \times \cdots \times \dfrac{2m-1}{2m+1} \leq \dfrac{1}{15}$,

$\dfrac{1}{2m+1} \leq \dfrac{1}{15}$,

$2m+1 \geq 15$,

$m \geq 7$。

答案　B

3. The function f is defined for all numbers x by $f(2x) = x^2 - 2x + 8$.

Quantity A	Quantity B
$f(6)$	12

A. Quantity A is greater.

B. Quantity B is greater.

C. The two quantities are equal.

D. The relationship cannot be determined from the information given.

题干翻译

对于所有 x，有函数 $f(2x) = x^2 - 2x + 8$。

解题思路

本题需要比较的是 $f(6)$ 与 12 的大小关系。

$2x = 6$，即 $x = 3$，

$f(6) = 3^2 - 2 \times 3 + 8 = 11 < 12$,

即数量 B 更大。

答案　B

2.9 函数图形

基本词汇

function 函数

平面直角坐标系可以用来描绘函数图像。要在平面直角坐标系中描绘函数图像，需要把每一个自变量 x 代入 $f(x)$，得出对应的点 (x, y)。以下是一些常见的基本函数图像。

（1）下图为函数 $y = -\dfrac{1}{2}x + 1$ 和 $y = x^2$ 的图像。

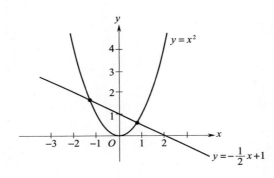

（2）下图为函数 $h(x) = |x|$，该函数 h 可以表达成一个分段形式：

$$h(x) = \begin{cases} x, & x \geq 0 \\ -x, & x < 0 \end{cases}$$

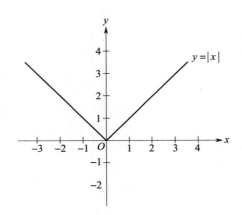

对于任何函数 $h(x)$ 和任何正数 c，有以下性质：
- 函数 $h(x)+c$ 的图像是 $h(x)$ 的图像向上平移 c 个单位。
- 函数 $h(x)-c$ 的图像是 $h(x)$ 的图像向下平移 c 个单位。
- 函数 $h(x+c)$ 的图像是 $h(x)$ 的图像向左平移 c 个单位。
- 函数 $h(x-c)$ 的图像是 $h(x)$ 的图像向右平移 c 个单位。

口诀：上加下减，左加右减。

例题

In the xy-plane, line k has x-intercept 2 and y-intercept 3. Line l is obtained by shifting line k to the right by 3 units and the downward by 2 units. What is the y-intercept of the line l?

A. $-\dfrac{11}{2}$ B. $-\dfrac{7}{2}$ C. 1   D. 3   E. $\dfrac{11}{2}$

题干翻译

在 xy 平面上，直线 k 的 x 轴截距为 2，y 轴截距为 3。直线 l 是通过将直线 k 向右移动 3 个单位，向下移动 2 个单位得到的。直线 l 的 y 轴截距是多少？

解题思路

根据题干条件"直线 k 的 x 轴截距为 2，y 轴截距为 3"可知，直线 k 经过 $(2,0)$ 和 $(0,3)$ 这两个点。

由此可以求得直线 k 的斜率 $=\dfrac{3-0}{0-2}=-\dfrac{3}{2}$，

直线 k 的表达式为 $y=-\dfrac{3}{2}x+3$。

根据题干条件"直线 l 是通过将直线 k 向右移动 3 个单位，向下移动 2 个单位得到的"和口诀"上加下减，左加右减"，

可知直线 l 的表达式为 $y=-\dfrac{3}{2}(x-3)+3-2=-\dfrac{3}{2}x+\dfrac{11}{2}$。

因此，直线 l 的 y 轴截距为：将 $x=0$ 代入表达式 $-\dfrac{3}{2}x+\dfrac{11}{2}$ 中，可得 $\dfrac{11}{2}$。

答案 E

第三章
CHAPTER

几何

几何部分主要考查平面几何、立体几何和解析几何。

注意，GRE 不考查证明过程。

几何部分主要考查 7 个知识点：
（1）直线和角
（2）多边形
（3）三角形
（4）四边形
（5）圆
（6）立体几何
（7）坐标几何

3.1 直线和角

基本词汇

- plane 平面
- polygon 多边形
- point 点
- line 线
- line segment 线段
- midpoint 中点
- angle 角
- degree 度
- opposite angle 对角
- vertical angle 对顶角
- congruent angle 全等角
- right angle 直角
- acute angle 锐角
- obtuse angle 钝角

平面几何主要考查平面图形的性质与关系，例如，角、三角形、其他多边形和圆。

当两条直线相交于一点，就形成了 4 个角。如下图所示，每一个角都有一个顶点 P，点 P 为两条直线的交点。在下图中，$\angle APC$ 和 $\angle BPD$ 叫作对角，或者对顶角。$\angle APD$ 与 $\angle BPC$ 也叫作对角，或者对顶角。对顶角的角度大小相等。角度大小相等的角叫作全等角。图中 4 个角的角度大小之和为 $360°$。

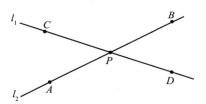

相交且组成 4 个全等角的两条直线互相垂直，4 个角大小均为 $90°$。角度大小为 $90°$ 的角为直角。直线 l_1 与直线 l_2 垂直可以记作 $l_1 \perp l_2$。

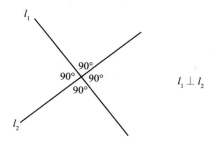

我们常用的标记直角的方法是在角的顶点处画一个小正方形。如下图所示，$\angle PON$ 是直角。

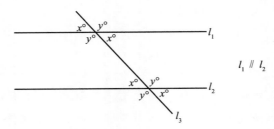

$ON \perp OP$

角度大小小于90°的角叫作锐角，角度大小在90°到180°之间的角叫作钝角。

同一平面内不相交的两条直线互相平行。如下图，直线 l_1 与直线 l_2 是平行的，我们可以记作 $l_1 \parallel l_2$。一条直线 l_3 与这两条直线相交，组成了 8 个角。注意，其中 4 个角的大小为 $x°$，剩下 4 个角的大小为 $y°$，$x+y=180$。

$l_1 \parallel l_2$

例题

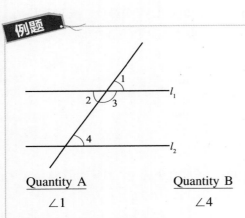

Quantity A　　　　Quantity B
∠1　　　　　　　∠4

A. Quantity A is greater.
B. Quantity B is greater.
C. The two quantities are equal.
D. The relationship cannot be determined from the information given.

解题思路
注意，GRE 数学图形不一定按比例绘制，因此如果题干并没有给出两条直线是否平行的信息，我们无法判断∠1 和∠4 的大小关系。

答案　D

3.2 多边形

> **基本词汇**
>
> side 边
> vertex 顶点
> convex polygon 凸多边形
> triangle 三角形
> quadrilateral 四边形
> pentagon 五边形
> hexagon 六边形
> heptagon 七边形
> octagon 八边形
> nonagon 九边形
> decagon 十边形
> regular polygon 正多边形
> perimeter 周长
> area 面积

概念

由 3 条以及更多条线段组成的封闭图形叫作多边形，这些线段叫作边，每条边在两个端点处与另外两条边相连，这些端点叫作顶点。注意，GRE 数学中的多边形都是凸多边形，即多边形的每个内角都小于 180°。如下图所示：

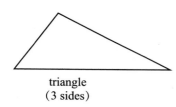
triangle
(3 sides)

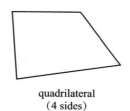
quadrilateral
(4 sides)

pentagon
(5 sides)

最简单的多边形是三角形。注意，一个四边形可以被分成两个三角形，一个五边形可以被分成 3 个三角形，如下图所示。

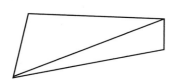

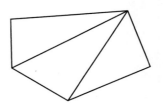

如果一个多边形有 n 条边，那么它就可以被分成 $n-2$ 个三角形。因为三角形的内角和是 180°，因此 n 边形的内角和就为 $(n-2) \times 180°$。比如，四边形的内角和为 $(4-2) \times 180° = 360°$。

如果一个多边形所有的边全等且所有的内角全等，那么这个多边形被称为正多边形。比如，对于正八边形而言，它的内角和为 $(8-2)\times 180°=1080°$，它的每个角都为 $\dfrac{1080°}{8}=135°$。

一个多边形的周长等于它每条边的长度之和，一个多边形的面积指的是多边形围起来的区域的面积。

例题

The sum of the measures of the interior angles of a pentagon is p degrees, and the sum of the measures of the interior angles of a hexagon is h degrees.

Quantity A　　　　Quantity B

$p+180$　　　　　h

A. Quantity A is greater.
B. Quantity B is greater.
C. The two quantities are equal.
D. The relationship cannot be determined from the information given.

题干翻译

五边形的内角之和是 p 度，六边形的内角之和是 h 度。

解题思路

本题需要我们比较 $p+180$ 和 h 的大小关系。

多边形内角和 =（边的个数 $n-2$）$\times 180$，

因此，这道题的关键在于知道 pentagon 和 hexagon 分别是几边形。

pentagon 是"五边形"，大家可以用美国的"五角大楼 the Pentagon"来联想记忆（同一个单词），hexagon 是"六边形"，大家可以将 hex 与 six 联想在一起来记。

五边形的内角和 $p=(5-2)\times 180=540$，

数量 A $=p+180=720$，

数量 B = 六边形的内角和 $h=(6-2)\times 180=720$，

数量 A 与数量 B 相等。

答案　C

3.3 三角形

基本词汇

equilateral triangle 等边三角形
isosceles triangle 等腰三角形
right triangle 直角三角形
hypotenuse 斜边
leg 直角边

opposite 对边
adjacent 邻边
congruent triangle 全等三角形
similar triangle 相似三角形

知识点

（1）每个三角形都有 3 条边和 3 个内角，三角形的内角和为 180°。三角形的一个外角等于其不相邻的两个内角之和。

（2）三条边能够构成三角形的条件：两边之和大于第三边，两边之差小于第三边。

（3）在同一个三角形中，大边对大角，大角对大边。

（4）三条边相等的三角形叫作等边三角形，等边三角形的每个内角都是 60°。

（5）至少有两条边相等的三角形叫作等腰三角形。如果一个三角形内有两条边相等，那么这两条边对应的角也相等，反之亦然。例如右图，△ABC 中∠A 和∠C 都是 50°，我们可以推出 $AB = BC$。因为三角形内角和为 180°，我们也可以推出 $\angle B = x = 180° - 50° - 50° = 80°$。

（6）如果三角形有一个内角为 90°，那么这个三角形叫作直角三角形。直角所对的边叫作斜边，剩下的两条边叫作直角边。如右图所示，在直角三角形 DEF 中，EF 是斜边，DE 和 DF 是直角边。根据勾股定理，在直角三角形中，斜边的平方等于两条直角边的平方和。因此在直角三角形 DEF 中，$EF^2 = DE^2 + DF^2$。

拓展性质：直角三角形斜边最长，斜边长度大于直角边长度。

勾股定理还可以用于非直角的三角形中：

- 如果三角形中较小的两条边的平方和小于第三边的平方，则该三角形为钝角三角形。
- 如果三角形中较小的两条边的平方和大于第三边的平方，则该三角形为锐角三角形。

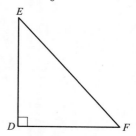

勾股定理还可以用来计算两种特殊的直角三角形的边长之比。一种是特殊的直角三角形——等腰直角三角形，根据勾股定理，等腰直角三角形三条边长度之比为 $1:1:\sqrt{2}$，如下图所示：

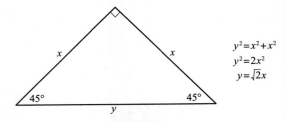

另一种是特殊的直角三角形，即 $30°-60°-90°$ 的直角三角形，它是一个等边三角形的一半，如下图所示：

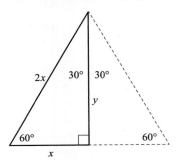

> **注意**
>
> 最短的边 x 是最长边 $2x$ 的一半，根据勾股定理，$x:y=1:\sqrt{3}$，因此这样的三角形三边长度之比为 $1:\sqrt{3}:2$。
>
> $$x^2+y^2=(2x)^2=4x^2$$
> $$y^2=3x^2$$
> $$y=\sqrt{3}x$$

(7) 三角函数

正弦 $\sin A = \dfrac{\angle A \text{ 的对边}}{\text{斜边}}$

余弦 $\cos A = \dfrac{\angle A \text{ 的邻边}}{\text{斜边}}$

正切 $\tan A = \dfrac{\angle A \text{ 的对边}}{\angle A \text{ 的邻边}}$

常用三角函数

$\sin 30° = \dfrac{1}{2}$ $\qquad\qquad$ $\sin 45° = \dfrac{\sqrt{2}}{2}$ $\qquad\qquad$ $\sin 60° = \dfrac{\sqrt{3}}{2}$

$\cos 30° = \dfrac{\sqrt{3}}{2}$ $\qquad\qquad$ $\cos 45° = \dfrac{\sqrt{2}}{2}$ $\qquad\qquad$ $\cos 60° = \dfrac{1}{2}$

$$\tan 30° = \frac{\sqrt{3}}{3} \qquad \tan 45° = 1 \qquad \tan 60° = \sqrt{3}$$

（8）三角形的面积 A 等于底乘高的一半。如下图所示，面积为 A，底边长为 b，底边对应的高为 h，则 $A = \frac{bh}{2}$。

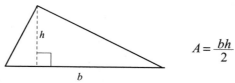

三角形任意一条边都可以用作底，底边上的高是一条从该底边所对的顶点到底边或底边延长线的垂线段。下图所示为三种不同三角形底边和底边上的高。

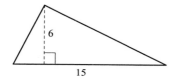  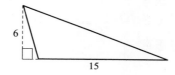

上面三种三角形，面积都为：$\frac{15 \times 6}{2} = 45$。

（9）全等三角形与相似三角形

如果两个三角形形状和面积大小相等，那么这两个三角形称为全等三角形。如果两个三角形内每个顶点所对应的角和每个角所对的边相等，那么这两个三角形全等。

以下 3 条性质可以帮我们来判断两个三角形是否全等。

- 如果一个三角形的三条边与另外一个三角形的三条边对应相等，那么这两个三角形全等。
- 如果一个三角形的两条边和它们所夹的角与另一个三角形的两条边和它们所夹的角都相等，那么这两个三角形全等。
- 如果一个三角形的两个角与另外一个三角形的两个角对应相等，且这两个角夹的边也对应相等，那么这两个三角形全等。

如果两个三角形形状相同但大小不同，那么这两个三角形被称为相似三角形。如果两个三角形每个顶点所对应的角相等或者说每个顶点所对应的边的比例相等，那么这两个三角形相似，这个比例被称为相似比。

如果 $\triangle ABC$ 与 $\triangle DEF$ 相似，那么 $\angle A$ 和 $\angle D$ 相等，$\angle B$ 和 $\angle E$ 相等，$\angle C$ 和 $\angle F$ 相等，如下图所示。字母的顺序代表相似的对应顺序。

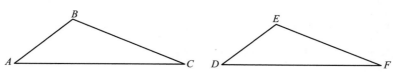

因为△ABC 与△DEF 相似，那么 $\frac{AB}{DE} = \frac{BC}{EF} = \frac{AC}{DF}$。

如果两个三角形相似，那么这两个三角形的面积比等于相似比的平方。

例题 01.

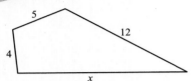

For the convex polygon above, which of the following intervals contains all possible values of x?

A. $1 < x < 17$ B. $3 < x < 21$ C. $4 < x < 12$ D. $4 < x < 21$ E. $7 < x < 21$

题干翻译

对于上面的凸多边形，下列哪个区间包含 x 的所有可能值？

解题思路

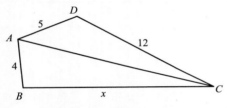

(为了方便讲解，将图形顶点标记 ABCD，如上图所示。)

连接 AC，

对于△ADC 而言，根据三角形两边之和大于第三边，两边之差小于第三边可知，$CD - AD < AC < CD + AD$，即 $7 < AC < 17$。

对于△ABC 而言，根据三角形两边之和大于第三边，两边之差小于第三边可知，AC（最小可能值）$- AB < BC < AC$（最大可能值）$+ AB$。由于 $7 < AC < 17$，因此 $7 - 4 < BC < 17 + 4$，即 $3 < x < 21$。

答案 B

例题 02.

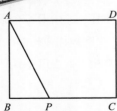

In the rectangle $ABCD$, $AB = 6$, $AD = 8$. If the point P is selected from segment BC at random, what is the probability that the length of segment AP is less than $3\sqrt{5}$?

A. $\dfrac{3}{8}$ B. $\dfrac{\sqrt{19}}{8}$ C. $\dfrac{3}{5}$ D. $\dfrac{3}{4}$ E. $\dfrac{3\sqrt{5}}{8}$

题干翻译

在矩形 $ABCD$ 中，$AB=6$，$AD=8$。如果在线段 BC 上随机选择点 P，线段 AP 的长度小于 $3\sqrt{5}$ 的概率是多少？

解题思路

考虑临界情况，$AP=3\sqrt{5}$。

对于直角 $\triangle ABP$，利用勾股定理 $AP^2=AB^2+BP^2$，即 $(3\sqrt{5})^2=6^2+BP^2$，

可得 $BP=3$，这意味着当点 B 和点 P 的距离为 3 时，AP 刚好为 $3\sqrt{5}$。

如果点 P 向左移动，AP 的长度变短；如果点 P 向右移动，AP 的长度变长。

因此，在长度为 8 的线段 BC 上，只要点 P 到点 B 的距离 <3，都能使得 $AP<3\sqrt{5}$。

即满足要求的概率 $=\dfrac{3}{8}$。

答案 A

例题 03

The lengths of the sides of triangle RST are 3, 4, and y. Which of the following inequalities specifies those values of y for which each angle measure of triangle RST is less than $90°$?

A. $1<y<4$ B. $1<y<5$ C. $2<y<5$ D. $\sqrt{7}<y<5$ E. $\sqrt{7}<y<6$

题干翻译

三角形 RST 的边长是 3，4 和 y。下列哪一个关于 y 的不等式能确定三角形 RST 的每个角都小于 $90°$？

解题思路

每个角都小于 $90°$ 的三角形为锐角三角形，即本题需要让我们确定能使得三角形 RST 为锐角三角形的 y 的取值。

根据勾股定理，如果三角形中较小的两条边的平方和大于第三边的平方，则该三角形为锐角三角形。由于本题不知道 y 的取值范围，我们需要分情况讨论。

情况 1：

$y\geqslant 4$，此时 y 为最长边。

$3^2+4^2>y^2$，

$25>y^2$，

$5>y$，并且 y 要满足 $y\geqslant 4$，

即 $4\leqslant y<5$。

情况2：

$y < 4$，此时4为最长边。

$3^2 + y^2 > 4^2$，

$y^2 > 7$，

$y > \sqrt{7}$，并且y要满足$y < 4$，

即$\sqrt{7} < y < 4$。

取情况1和情况2的并集，可得$\sqrt{7} < y < 5$。

答案　D

例题 04.

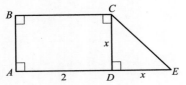

In the figure above, the area of quadrilateral region $ABCE$ is $\frac{5}{2}$. What is the value of x?

A. $\frac{3}{4}$　　　　B. 1　　　　C. $\frac{5}{4}$　　　　D. $\frac{3}{2}$　　　　E. 2

题干翻译

在上图中，四边形区域$ABCE$的面积是$\frac{5}{2}$。x的值是多少？

解题思路

$ABCE$的面积 = 矩形$ABCD$的面积 + 三角形CDE的面积，

$$= 2x + \frac{1}{2}x^2 = \frac{5}{2}。$$

$x^2 + 4x - 5 = 0$，

$(x+5)(x-1) = 0$，

$x = 1$（x为边长，必须>0，因此排除$x = -5$的情况）。

答案　B

例题 05.

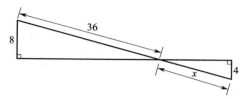

In the figure above, x equals

A. $4\frac{1}{2}$ B. 8 C. 12 D. 16 E. 18

解题思路

从上图可以看出，这两个三角形3个对应的角相等，因此这两个三角形为相似三角形。

因此，$\frac{36}{8} = \frac{x}{4}$，

$x = 18$。

答案 E

练习

1. The sum of the lengths of two sides of isosceles triangle T is 10, and T has a side of length 5.

Quantity A	Quantity B
The perimeter of T	15

   A. Quantity A is greater.
   B. Quantity B is greater.
   C. The two quantities are equal.
   D. The relationship cannot be determined from the information given.

2.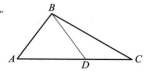

In the figure shown, $AB = BD = DC$ and the degree measure of angle ABD is 80.

Quantity A	Quantity B
The degree measure of angle DBC	30

A. Quantity A is greater.

B. Quantity B is greater.

C. The two quantities are equal.

D. The relationship cannot be determined from the information given.

3.

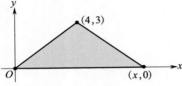

The area of the shaded triangular region is 12.

Quantity A	Quantity B
x	8

A. Quantity A is greater.

B. Quantity B is greater.

C. The two quantities are equal.

D. The relationship cannot be determined from the information given.

4.

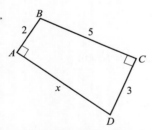

In quadrilateral ABCD shown, what is the value of x?

Give your answer to the nearest 0.1.

5.

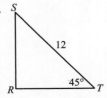

Quantity A	Quantity B
The area of triangular region RST	36

A. Quantity A is greater.

B. Quantity B is greater.

C. The two quantities are equal.

D. The relationship cannot be determined from the information given.

6.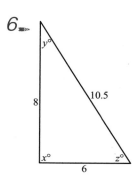

Quantity A	Quantity B
x	$y + z$

A. Quantity A is greater.

B. Quantity B is greater.

C. The two quantities are equal.

D. The relationship cannot be determined from the information given.

7.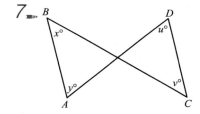

$v < u, \ x < y$

Quantity A	Quantity B
BC	AD

A. Quantity A is greater.

B. Quantity B is greater.

C. The two quantities are equal.

D. The relationship cannot be determined from the information given.

8.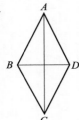

Quantity A	Quantity B
The sum of the two diagonals of quadrilateral ABCD	The perimeter of quadrilateral ABCD

A. Quantity A is greater.

B. Quantity B is greater.

C. The two quantities are equal.

D. The relationship cannot be determined from the information given.

答案及解析

1. The sum of the lengths of two sides of isosceles triangle T is 10, and T has a side of length 5.

Quantity A	Quantity B
The perimeter of T	15

A. Quantity A is greater.

B. Quantity B is greater.

C. The two quantities are equal.

D. The relationship cannot be determined from the information given.

题干翻译

等腰三角形 T 的两条边的长度之和是 10，T 有一条边的长度为 5。

解题思路

本题需要我们比较三角形 T 的周长和 15 的大小关系。

如果这个等腰三角形是等边三角形，那么三角形 T 的周长 = 15，数量 A 与数量 B 相等。

如果这个等腰三角形不是等边三角形，那么数量 A 与数量 B 不相等：当两条腰的长度均为 5 时，0 < 第三边长度 < 10，10 < 三角形 T 的周长 < 20。

答案 D

2.

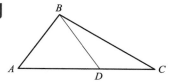

In the figure shown, $AB = BD = DC$ and the degree measure of angle ABD is 80.

Quantity A	Quantity B
The degree measure of angle DBC	30

A. Quantity A is greater.

B. Quantity B is greater.

C. The two quantities are equal.

D. The relationship cannot be determined from the information given.

题干翻译

在上图中,$AB = BD = DC$,$\angle ABD$ 的度数为 $80°$。

解题思路

本题需要我们比较 $\angle DBC$ 的度数与 $30°$ 的大小关系。

$AB = BD$,因此 △ABD 为等腰三角形,$BD = DC$,因此 △BDC 为等腰三角形。

在等腰三角形 ABD 中,$\angle ABD = 80°$,那么 $\angle BAD$ 和 $\angle BDA$ 均为 $50°$。

在等腰三角形 BDC 中 $\angle BDC = 130°$,$\angle DBC = \angle BCD = 25°(<30°)$。

即数量 B 更大。

答案 B

3.

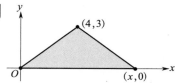

The area of the shaded triangular region is 12.

Quantity A	Quantity B
x	8

A. Quantity A is greater.

B. Quantity B is greater.

C. The two quantities are equal.

D. The relationship cannot be determined from the information given.

题干翻译

阴影三角形区域的面积是 12。

解题思路

本题需要我们比较 x 与 8 的大小关系。

根据图形可以看出三角形的底边长度为 x，高度为 3。

三角形的面积 $= \frac{1}{2}$ 底 × 高 $= \frac{1}{2} x \times 3 = 12$，

$x = 8$，

即数量 A 与数量 B 一样大。

答案　C

4.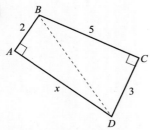

In quadrilateral *ABCD* shown, what is the value of x? Give your answer to the nearest 0.1.

题干翻译

在上图的四边形 *ABCD* 中，x 的值是多少？

解题思路

连接 *BD*。

在直角三角形 *BCD* 中，$BD^2 = BC^2 + CD^2 = 5^2 + 3^2 = 34$，

在直角三角形 *BAD* 中，$BD^2 = AB^2 + AD^2 = 2^2 + x^2 = 34$，

$x^2 = 30$，

$x \approx 5.5$。

答案　5.5

5.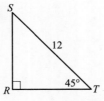

Quantity A	Quantity B
The area of triangular region *RST*	36

A. Quantity A is greater.

B. Quantity B is greater.

C. The two quantities are equal.

D. The relationship cannot be determined from the information given.

解题思路

本题需要我们比较三角形 SRT 的面积与 36 的大小关系。

根据上图可知，三角形为等腰直角三角形（$45°$，$45°$，$90°$）。

$SR = RT = ST \times \sin 45° = 12 \times \dfrac{\sqrt{2}}{2} = 6\sqrt{2}$，

因此，三角形 RST 的面积 $= \dfrac{1}{2}$ 底 $\times$ 高 $= \dfrac{1}{2} \times (6\sqrt{2})^2 = 36$，

故数量 A 与数量 B 大小相等。

答案　C

6.

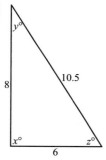

Quantity A	Quantity B
x	$y + z$

A. Quantity A is greater.

B. Quantity B is greater.

C. The two quantities are equal.

D. The relationship cannot be determined from the information given.

解题思路

本题需要我们比较 x 与 $y + z$ 的大小关系。

根据图形信息可以得到：$8^2 + 6^2 < 10.5^2$，

因此该三角形为钝角三角形，因此 $x > 90 > y + z$，

故数量 A 更大。

答案　A

7.

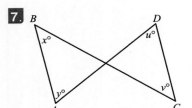

$v < u$, $x < y$

Quantity A	Quantity B
BC	AD

A. Quantity A is greater.

B. Quantity B is greater.

C. The two quantities are equal.

D. The relationship cannot be determined from the information given.

解题思路

本题要求我们比较 BC 与 AD 的大小关系。

同一个三角形内大角对大边，大边对大角。

设交点为 O。

在三角形 DOC 中，根据 $v < u$ 可知，$OD < OC$，

在三角形 AOB 中，根据 $x < y$ 可知，$OA < OB$，

因此 $OA + OD < OB + OC$，

即 $AD < BC$，

故数量 A 更大。

答案 A

##

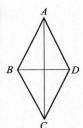

Quantity A	Quantity B
The sum of the two diagonals of quadrilateral ABCD	The perimeter of quadrilateral ABCD

A. Quantity A is greater.

B. Quantity B is greater.

C. The two quantities are equal.

D. The relationship cannot be determined from the information given.

解题思路

本题要求我们比较四边形 ABCD 两条对角线的和与四边形 ABCD 的周长的大小关系。

根据三角形两边之和大于第三边可得：

式子 1：$AB + AD > BD$；

式子 2：$BC + CD > BD$；

式子 3：$AB + BC > AC$；

式子 4：$AD + DC > AC$；

式子 1 + 式子 2 + 式子 3 + 式子 4

$= 2(AB + AD + BC + CD) > 2(AC + BD)$，

因此 $AB + AD + BC + CD > AC + BD$，

即四边形 ABCD 的周长 > 四边形 ABCD 两条对角线的和，

故数量 B 更大。

答案 B

3.4 四边形

基本词汇

rectangle 矩形　　　parallelogram 平行四边形　　　rhombus 菱形
square 正方形　　　trapezoid 梯形　　　diagonal 对角线

概念

每个四边形都有 4 条边和 4 个内角，4 个内角之和为 360°。
GRE 中常考的四边形有以下几种。

1. 矩形

如果一个四边形的 4 个角都是直角，那么这个四边形叫作矩形。矩形的两组对边平行且相等，两条对角线相等。

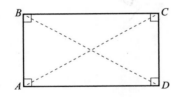

$AB \parallel CD$ and $AD \parallel BC$
$AB=CD$ and $AD=BC$
$AC=BD$

矩形的周长 =（长 + 宽）×2。
矩形的面积 = 长 × 宽。
矩形对角线的长度 = $\sqrt{长^2 + 宽^2}$。

2. 正方形

4 条边全等的矩形叫作正方形。
正方形的周长 = 边长 ×4。
正方形的面积 = 边长^2。

3. 平行四边形

① 平行四边形的判定依据

从边看：两组对边分别平行的四边形为平行四边形；
　　　　两组对边分别相等的四边形为平行四边形；
　　　　一组对边平行且相等的四边形为平行四边形。
从角看：两组对角分别相等的四边形为平行四边形。
　　　　从对角线看：对角线互相平分的四边形为平行四边形。

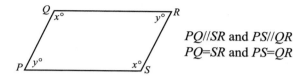

PQ//SR and PS//QR
PQ=SR and PS=QR

②平行四边形的周长＝(底边＋侧边)×2。
③平行四边形的面积＝底×高。
对于所有的平行四边形，包含矩形和正方形，面积 A 等于底边 b 乘底边对应的高 h。
即 $A = bh$。
平行四边形的任意一边都可以做底，底对应的高是从底边上任意一点到对边或者对边延长线的垂线段。下面是一个矩形和一个平行四边形的面积计算：

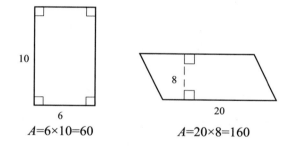

$A=6×10=60$　　　　$A=20×8=160$

4. 梯形

如果一个四边形只有一组对边平行，那么这个四边形叫作梯形。

梯形的面积 A 等于平行的两条边 b_1，b_2 的和的一半，乘其对应的高 h。

梯形的面积 $= \frac{1}{2}$(上底＋下底)×高，

即 $A = \frac{1}{2}(b_1 + b_2)h$。

比如，两个平行的边长度为 10 和 18，高为 7.5 的梯形，

面积 $A = \frac{1}{2}(10 + 18) × 7.5 = 105$。

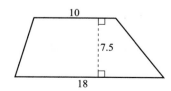

5. 菱形

菱形的判定依据有以下几条。

①四条边都相等的四边形为菱形。

②对角线互相垂直的平行四边形为菱形。

③有一组邻边相等的平行四边形为菱形。

拓展性质：

如果一个四边形的对角线互相垂直，那么这个四边形的面积 = 对角线长度的乘积 ÷ 2。

6. 正六边形

①正六边形是由 6 个等边三角形组成的。

②正六边形的面积 = 等边三角形的面积 × 6。

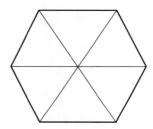

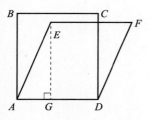

Square *ABCD* and parallelogram *AEFD* lie in the same plane, and $AB = AE$.

Quantity A	Quantity B
The area of region *ABCD*	The area of region *AEFD*

A. Quantity A is greater.

B. Quantity B is greater.

C. The two quantities are equal.

D. The relationship cannot be determined from the information given.

题干翻译

正方形 $ABCD$ 和平行四边形 $AEFD$ 位于同一平面，$AB = AE$。

解题思路

本题需要我们比较区域 $ABCD$ 的面积和 $AEFD$ 的面积。

正方形 $ABCD$ 和平行四边形 $AEFD$ 底相同，均为 AD。

正方形 $ABCD$ 的高 CD 大于平行四边形 $AEFD$ 的高（图形为平面图形，经过 E 点作平行四边形 $AEFD$ 的高 EG，而 EG 是直角边，它会小于对应斜边 AE（$=AB$），

而四边形的面积＝底×高，

因此，$ABCD$ 的面积比 $AEFD$ 的面积大。

答案　A

The length and width of rectangle R is x and y, respectively, and the diagonal of R has length 12.

<u>Quantity A</u>　　　　<u>Quantity B</u>

$(x+y)^2$　　　　　　144

A. Quantity A is greater.

B. Quantity B is greater.

C. The two quantities are equal.

D. The relationship cannot be determined from the information given.

题干翻译

矩形 R 的长度和宽度分别为 x 和 y，矩形 R 的对角线长度为 12。

解题思路

本题需要我们比较 $(x+y)^2$ 与 144 的大小关系。

矩形对角线的长度 $= \sqrt{长^2+宽^2} = \sqrt{x^2+y^2} = 12$，

$x^2 + y^2 = 144$，

$(x+y)^2 = x^2 + y^2 + 2xy = 144 + 2xy$。

由于 x，y 为矩形的边长，边长 > 0，因此 $144 + 2xy > 144$，

故数量 A 更大。

答案　A

例题 03.

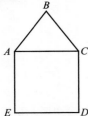

In the figure above, the area of equilateral triangle *ABC* is $25\sqrt{3}$. What is the area of square *ACDE*?

题干翻译

上图中，等边三角形 *ABC* 的面积是 $25\sqrt{3}$。正方形 *ACDE* 的面积是多少？

解题思路

等边三角形 *ABC* 的面积 $=\frac{1}{2}底\times高=\frac{1}{2}AC\times AB\times\sin60°$（等边三角形的每个内角为 $60°$），

$\frac{1}{2}AC^2\times\frac{\sqrt{3}}{2}=25\sqrt{3}$（等边三角形 3 条边相等 $AB=AC=BC$），

$AC^2=100$，$AC=10$，

正方形 *ACDE* 的面积 $=边长^2=AC^2=100$。

答案 100

例题 04.

Let S and T be trapezoids, each of which has two nonparallel opposite sides that are congruent. The bases of S have lengths 8 and 18, the bases of T have lengths 8 and 26, and the heights of S and T are equal. If the perimeter of S is 52, what is the perimeter of T?

A. 60　　　　B. 64　　　　C. 68　　　　D. 72　　　　E. 76

题干翻译

S 和 T 是梯形，每个梯形都有两条不平行的全等对边。梯形 S 的底部长度为 8 和 18，梯形 T 的底部长度为 8 和 26，梯形 S 和梯形 T 的高度相等。如果梯形 S 的周长是 52，那么梯形 T 的周长是多少？

解题思路

设梯形 S 的两条不平行的全等对边长度为 x。

根据"梯形 S 的周长为 52"这一条件可知：$8+26+2x=52$，$x=13$。

设梯形 S 的高为 h。

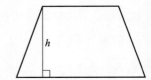

根据勾股定理可知：$h^2 + \left(\dfrac{18-8}{2}\right)^2 = 13^2$，

$h^2 = 144$，

$h = 12$。

设梯形 T 的两条不平行的全等对边长度为 y，

根据勾股定理可知：$h^2 + \left(\dfrac{26-8}{2}\right)^2 = y^2$，

根据题干信息"两个梯形的高度相等"可知梯形 T 的高 $h = 12$，代入上述等式，得到：

$12^2 + 9^2 = y^2$，

$y^2 = 225$，

$y = 15$。

因此，梯形 T 的周长 $= 15 \times 2 + 8 + 26 = 64$。

答案 B

例题 05.

The figure shows a regular hexagon inscribed in a circle with circumference 8π. Approximately what is the sum of the areas of the shaded regions?

A. 4　　　　　B. 9　　　　　C. 35　　　　　D. 42　　　　　E. 43

题干翻译

该图显示一个正六边形内接在一个周长为 8π 的圆上。阴影区域的面积之和大约是多少？

解题思路

根据"圆的周长为 8π"可知，圆的半径 $r = \dfrac{8\pi}{2\pi} = 4$。

连接圆心和正六边形的顶点构成 6 个等边三角形（因为圆的半径相等，且半径夹的角 $= \dfrac{360°}{6} = 60°$），

等边三角形的边长 = 圆的半径 = 4，

> 等边三角形的面积 $= \frac{1}{2} r \times r \times \sin 60° = 4\sqrt{3}$,
>
> 正六边形的面积和 $= 6 \times$ 等边三角形的面积 $= 24\sqrt{3}$,
>
> 阴影部分的面积 $=$ 圆的面积 $-$ 正六边形的面积 $= 16\pi - 24\sqrt{3} \approx 9$。
>
> 答案　B

练 习

1. A rectangle with perimeter 60 has 2 sides of length $2x - 8$ and 2 sides of length $x + 2$.

Quantity A	Quantity B
$2x - 8$	$x + 2$

   A. Quantity A is greater.
   B. Quantity B is greater.
   C. The two quantities are equal.
   D. The relationship cannot be determined from the information given.

2.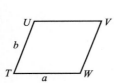

   PQRS is a rectangle and TUVW is a parallelogram, with side lengths as indicated. The measure of angle UTW is less than 90 degrees.

Quantity A	Quantity B
The area of region PQRS	The area of region TUVW

   A. Quantity A is greater.
   B. Quantity B is greater.
   C. The two quantities are equal.
   D. The relationship cannot be determined from the information given.

3. What is the ratio of the area of a square region with diagonal 10 to the area of a square region with diagonal 20?

   A. 1∶4　　B. 1∶2　　C. 1∶$\sqrt{2}$　　D. 2∶1　　E. 4∶1

4.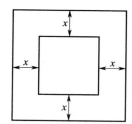

The figure shows a square patio surrounded by a walkway of width x meters. If the area of the walkway is 132 square meters and the width of the patio is 5 meters greater than the width of the walkway, what is the area of the patio in square meters?

A. 56   B. 64   C. 68   D. 81   E. 100

5.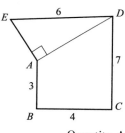

Quantity A	Quantity B
The perimeter of the quadrilateral $ABCD$	20

A. Quantity A is greater.
B. Quantity B is greater.
C. The two quantities are equal.
D. The relationship cannot be determined from the information given.

1. A rectangle with perimeter 60 has 2 sides of length $2x-8$ and 2 sides of length $x+2$.

Quantity A	Quantity B
$2x-8$	$x+2$

A. Quantity A is greater.
B. Quantity B is greater.
C. The two quantities are equal.
D. The relationship cannot be determined from the information given.

题干翻译

周长为 60 的矩形有两条长度为 $2x-8$ 的边和两条长度为 $x+2$ 的边。

解题思路

本题需要我们比较 $2x-8$ 与 $x+2$ 的大小关系。

矩形的周长 =（长 + 宽）×2，

$(2x-8+x+2) \times 2 = 60$，

$3x - 6 = 30$，

$x = 12$，

$2x - 8 = 16$，

$x + 2 = 14 < 16$，

故数量 A 更大。

答案 A

2.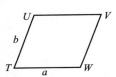

PQRS is a rectangle and TUVW is a parallelogram, with side lengths as indicated. The measure of angle UTW is less than 90 degrees.

Quantity A	Quantity B
The area of region PQRS	The area of region TUVW

A. Quantity A is greater.

B. Quantity B is greater.

C. The two quantities are equal.

D. The relationship cannot be determined from the information given.

题干翻译

PQRS 是一个矩形，TUVW 是一个平行四边形，边长如图所示。∠UTW < 90°。

解题思路

本题需要我们比较区域 PQRS 与区域 TUVW 的面积大小。

矩形 PQRS 的面积 = 底 × 高 = ab，

平行四边形 TUVW 的面积 = 底 × 高 = a × 高。

根据题干条件 ∠UTW < 90° 可知，平行四边形 TUVW 底边 a 对应的高 < b（直角边 < 斜边），

因此，矩形 PQRS 的面积 > 平行四边形 TUVW 的面积，
故数量 A 更大。

答案 A

3. What is the ratio of the area of a square region with diagonal 10 to the area of a square region with diagonal 20?

A. $1:4$ B. $1:2$ C. $1:\sqrt{2}$ D. $2:1$ E. $4:1$

题干翻译

对角线为 10 的正方形区域的面积与对角线为 20 的正方形区域的面积之比是多少？

解题思路

我们把对角线为 10 的正方形称为正方形 1，对角线为 20 的正方形称为正方形 2。

相似图形的面积比 = 边长比²，

$\dfrac{\text{正方形 1 的对角线}}{\text{正方形 2 的对角线}} = \dfrac{\text{正方形 1 的边长}}{\text{正方形 2 的边长}} = \dfrac{10}{20} = \dfrac{1}{2}$，

$\dfrac{\text{正方形 1 的面积}}{\text{正方形 2 的面积}} = \text{边长比}^2 = \left(\dfrac{1}{2}\right)^2 = \dfrac{1}{4}$。

答案 A

4.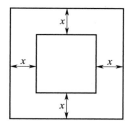

The figure shows a square patio surrounded by a walkway of width x meters. If the area of the walkway is 132 square meters and the width of the patio is 5 meters greater than the width of the walkway, what is the area of the patio in square meters?

A. 56 B. 64 C. 68 D. 81 E. 100

题干翻译

该图显示一个正方形的平台，被一条宽 x 米的走道所围绕。如果走道的面积是 132 平方米，平台的宽度比走道的宽度大 5 米，请问平台的面积是多少平方米？

解题思路

因为平台是正方形，走道的宽度均为 x 米，因此整个图形就是一个大正方形。

因为走道宽 x 米，所以平台的宽度为 $x+5$ 米，整个图形的边长为 $x+5+2x=3x+5$ 米。

走道面积 = 整个图形的面积 – 正方形平台的面积，

$132 = (3x+5)^2 - (x+5)^2$，

$x = 3$，

因此，平台面积 = $(x+5)^2 = 64$ 平方米。

答案 B

5.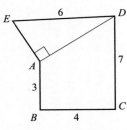

Quantity A	Quantity B
The perimeter of the quadrilateral ABCD	20

A. Quantity A is greater.

B. Quantity B is greater.

C. The two quantities are equal.

D. The relationship cannot be determined from the information given.

解题思路

本题需要我们比较四边形 ABCD 的周长和 20 的大小。

如图所示，AD 为直角三角形的直角边，因此 AD < 斜边 ED(=6)，

四边形 ABCD 的周长 = AB + BC + CD + AD = 3 + 4 + 7 + AD < 14 + ED(=20)，

故数量 B 更大。

答案 B

3.5 圆

> **基本词汇**
>
> circle 圆 　　circumference 周长 　　sector 扇形
> center 圆心 　　arc 弧 　　tangent 切线
> radius 半径 　　shorter arc 劣弧 　　point of tangency 切点
> diameter 直径 　　longer arc 优弧 　　be inscribed in 内接于
> congruent circle 全等圆 　　central angle 圆心角 　　be circumscribed about 外接于
> chord 弦 　　inscribed angle 圆周角 　　concentric circles 同心圆

概念

已知平面内一点 O 和一个正数 r，在该平面内与点 O 的距离为 r 的点的集合叫作圆。点 O 叫作圆的圆心，距离 r 叫作圆的半径。圆的直径是圆的半径的两倍。两个半径相同的圆叫作全等圆。连接圆上任意两点的线段叫作弦。连接圆上的任意一点和圆心形成的线段叫作半径，通过圆心的弦叫作直径，直径是最长的弦。如右图所示，O 为圆心，r 为半径，PQ 为弦，ST 为直径。

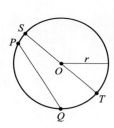

考点 2　圆的周长和面积

绕圆一周的距离叫作圆的周长，圆的周长 C 等于直径与 π 的乘积，π 的值大约为 3.14。
即 $C = \pi d$。
对于半径为 r 的圆来说，它的周长为半径的 2 倍与 π 的乘积。
即 $C = \pi \times 2r = 2\pi r$，
圆的面积 $= \pi r^2$。

考点 2　弧、圆周角和圆心角、扇形、切线

在圆上取任意两个点，这两个点以及两点之间圆上所有的点构成了弧。圆上的任意两个端点都可以看作两条弧的端点。通常我们用 3 个点来定义一条弧以避免模糊性。如下图所示，弧 ABC 是 AC 之间的劣弧，弧 ADC 是 AC 之间的优弧。

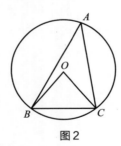

图1

圆心角是顶点在圆心的角。弧的两个端点所对应的半径交于圆心从而形成的角叫作弧对应的圆心角，弧度的大小就是其对应的圆心角的大小。如上图中的∠AOC，就是弧ABC对应的圆心角。我们可以将一整个圆看作一个大小为360°的弧。如上图所示，弧ABC的大小为50°，弧ADC的大小为310°。

弧长 = $\dfrac{\text{弧所对应的圆心角的角度}}{360°}$ × 圆的周长。

上图中弧ABC的长度 = $\dfrac{\text{弧}ABC\text{所对应的圆心角的角度}}{360°}$ × 圆的周长

$= \dfrac{50°}{360°} \times 2\pi \times 5$

$= \dfrac{25\pi}{18}$

圆周角：顶点在圆上的角，如下图∠BAC就是劣弧BC对应的圆周角。

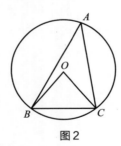

图2

关于圆周角的性质：
①直径对应的圆周角为90°，如下图∠ACB。
②同一段弧/弦对应的圆周角是圆心角的一半。

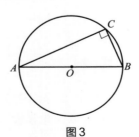

图3

扇形是圆中由弧和两条半径构成的区域。在图1中，弧 ABC 和两条虚线构成的区域即为圆心角为 $50°$ 的扇形。

扇形面积 = $\dfrac{\text{弧所对应的圆心角的角度}}{360°} \times$ 圆的面积。

图1中扇形 AOC 的面积 = $\dfrac{\text{弧所对应的圆心角的角度}}{360°} \times$ 圆的面积

$= \dfrac{50°}{360°} \times \pi \times 5^2$

$= \dfrac{125\pi}{36}$

如果一条直线与圆仅交于一点，那么这条直线成为圆的切线，相交的点成为切点。如下图的 P 点。如果一条直线与圆相切，那么切点对应的半径与切线垂直。

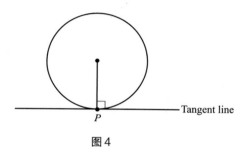

图4

考点3 同心圆

两个或者更多圆心相同的圆叫作同心圆，如下图所示。

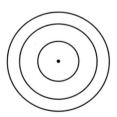

同心圆常考环形面积，环形面积 = 大圆面积 – 小圆面积。

考点4 圆的内接图形和外切图形

如果一个多边形的所有顶点都在圆上，那么这个多边形内接于该圆，或者说这个圆外接于该多边形。如下图所示，$\triangle RST$ 和 $\triangle XYZ$ 分别内接于圆 O 和圆 W。

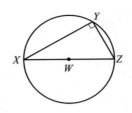

如果内接三角形的一条边为圆的直径，如上图中的△XYZ，那么这个三角形是直角三角形。如果一个内切三角形是直角三角形，那么该三角形其中一条边为圆的直径。

如果一个多边形的每条边都与圆相切，那么这个多边形外切于该圆，或者说该圆内切于该多边形。如下图所示，四边形 ABCD 外切于圆 O。

圆的内接正方形的对角线和圆的直径重合，所以圆的直径 = 内接正方形的对角线；

圆的外切正方形的边长和圆的直径相等，所以圆的直径 = 外切正方形的边长。

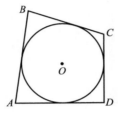
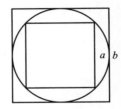

其他性质：内接于圆内的四边形对角互补。

考点5 圆的方程

圆心为 (a, b)，半径为 r 的圆的方程为：$(x-a)^2 + (y-b)^2 = r^2$。

例题 01.

$r > 0$

Quantity A	Quantity B
The area of a circular region with radius r	The area of a circular region with radius r^2

A. Quantity A is greater.
B. Quantity B is greater.
C. The two quantities are equal.
D. The relationship cannot be determined from the information given.

解题思路

本题需要我们比较半径为 r 的圆的面积和半径为 r^2 的圆的面积大小关系。

半径为 r 的圆的面积 $= \pi r^2$，

半径为 r^2 的圆的面积 $= \pi r^4$。

当 $0 < r < 1$ 时，$\pi r^2 > \pi r^4$，

当 $r = 1$ 时，$\pi r^2 = \pi r^4$，

当 $r > 1$ 时，$\pi r^2 < \pi r^4$。

因此，数量 A 与数量 B 的大小无法确定。

答案 D

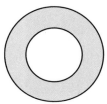

In the figure shown, if the area of the shaded region is 3 times the area of the smaller circular region, then the circumference of the larger circle is how many times the circumference of the smaller circle?

A. 4　　　　B. 3　　　　C. 2　　　　D. $\sqrt{3}$　　　　E. $\sqrt{2}$

题干翻译

如图所示，如果阴影区域的面积是较小圆形区域面积的 3 倍，那么较大圆形的周长是较小圆形的周长的多少倍？

解题思路

阴影部分的面积 = 大圆的面积 − 小圆的面积。

设大圆的半径为 R，小圆的半径为 r，因此阴影部分的面积 $= \pi R^2 - \pi r^2 =$ 小圆面积的 3 倍 $= 3\pi r^2$。

$\pi R^2 = 4\pi r^2$，

$R = 2r$，

所以，大圆的周长为 $2\pi R = 4\pi r$，小圆的周长为 $2\pi r$，

由此看出大圆的周长是小圆的周长的 2 倍。

答案 C

例题 03.

Triangle RST is inscribed in the circle with circumference of 20π and $\angle R > 90°$.

Quantity A	Quantity B
ST	20

A. Quantity A is greater.

B. Quantity B is greater.

C. The two quantities are equal.

D. The relationship cannot be determined from the information given.

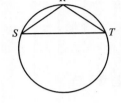

题干翻译

三角形 RST 内接于周长为 20π 的圆中，且 $\angle R > 90°$。

解题思路

本题需要我们比较 ST 和 20 的大小。

由题干可知，圆的周长为 20π，因此，圆的直径为 20。

$\angle R > 90°$，因此 ST 不是直径（直径对应的圆周角为 $90°$），而直径是圆中最长的弦，所以 $ST < 20$。

答案 B

例题 04.

A is the center of the circle and $\angle CAB = 50°$.

Quantity A	Quantity B
AB	BC

A. Quantity A is greater.

B. Quantity B is greater.

C. The two quantities are equal.

D. The relationship cannot be determined from the information given.

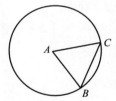

题干翻译

A 为圆的圆心，且 $\angle CAB = 50°$。

解题思路

根据题干条件 $\angle CAB = 50°$，且三角形 ABC 为腰长为 r 的等腰三角形，可知 $\angle ACB = 65°$。

同一三角形内大角对大边，

因此，$AB > BC$。

答案 A

例题 05.

In the figure shown, O is the center of the circle. If minor arc AB has length 2π, what is the area of triangular region AOB?

A. 8　　　B. 16　　　C. 4π　　　D. 8π　　　E. 16π

题干翻译

在图中，O 是圆心。如果劣弧 AB 的长度为 2π，那么三角形区域 AOB 的面积是多少？

解题思路

弧长 = $\dfrac{\text{弧所对应的圆心角的角度}}{360°}$ × 圆的周长，

因此，劣弧 AB 的长度 = $\dfrac{90°}{360°} \times 2\pi r = 2\pi$，

$r = 4$，

△AOB 面积 = $\dfrac{1}{2} \times 4 \times 4 = 8$。

答案　A

例题 06.

Quadrilateral $ABCD$ is inscribed in the circle shown. What is the value of $x + y$?

(Note: The measure of an angle inscribed in a circle is equal to half the measure of the central angle that subtends the same arc.)

A. 20　　　B. 22.5　　　C. 36　　　D. 52.5　　　E. 56

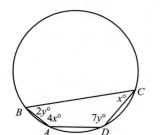

题干翻译

四边形 $ABCD$ 内接于右图所示的圆中。$x + y$ 的值是多少？

(注：圆内同一圆弧对应的圆周角是圆心角的一半。)

解题思路

内接于圆内的四边形对角互补。

因此，$2y + 7y = 180$，可得 $y = 20$，

$4x + x = 180$，可得 $x = 36$，

$x + y = 56$。

答案　E

例题 07.

In the figure shown, a circle is inscribed in a square with side b and a square with side a is inscribed in the circle. If $a = 4$, what is the area of the square with side b?

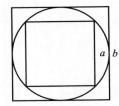

题干翻译

图中，一个圆内切一个边为 b 的正方形，一个边为 a 的正方形内接一个圆。如果 $a = 4$，边长为 b 的正方形的面积是多少？

解题思路

圆的直径 = 内接正方形的对角线 = $\sqrt{2}a = 4\sqrt{2}$，

圆的直径 = 外切正方形的边长 = $b = 4\sqrt{2}$，

因此，边长为 b 的正方形的面积 = $(4\sqrt{2})^2 = 32$。

答案 32

例题 08.

In the xy-plane, a circle with radius 4 has its center at the point $(-1, 2)$. Which of the following is an equation of the circle?

A. $(x-1)^2 + (y+2)^2 = 2$
B. $(x-1)^2 + (y+2)^2 = 4$
C. $(x-1)^2 + (y+2)^2 = 16$
D. $(x+1)^2 + (y-2)^2 = 4$
E. $(x+1)^2 + (y-2)^2 = 16$

题干翻译

在 xy 平面上，下列哪一项是半径为 4、圆心在点 $(-1, 2)$ 上的圆的方程？

解题思路

圆心为 (a, b)，半径为 r 的圆的方程为：$(x-a)^2 + (y-b)^2 = r^2$，

因此，该圆的方程为：$(x+1)^2 + (y-2)^2 = 4^2 = 16$。

答案 E

练 习

1. The area of a circular region with radius r is 5π.

Quantity A	Quantity B
The area of a circular region with radius $3r$	45π

A. Quantity A is greater.

B. Quantity B is greater.

C. The two quantities are equal.

D. The relationship cannot be determined from the information given.

2. In the xy-plane, which of the following points is NOT on the circle that has center $(0, 0)$ and radius 5?

A. $(-5, 0)$ B. $(-3, -4)$ C. $(1, -2\sqrt{6})$ D. $(3, 3\sqrt{3})$ E. $(\sqrt{21}, 2)$

3. The two circles in the figure shown both have center O. The square is inscribed in the larger circle and circumscribed about the smaller circle. If x is the radius of the larger circle, what is the area of the smaller circle, in terms of x?

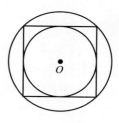

A. $\dfrac{3\pi x^2}{4}$ B. $\dfrac{2\pi x^2}{3}$ C. $\dfrac{\pi x^2}{2}$ D. $\dfrac{\pi x^2}{3}$

E. It cannot be determined from the information given.

4. Two concentric circles have diameters 6 and 10. What is the area of the region between the two circles?

A. 64π B. 36π C. 25π D. 16π E. 4π

5. The area of the shaded portion of the circular region with center O is 16π.

Quantity A	Quantity B
The radius of the circle	5.5

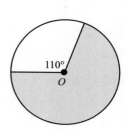

A. Quantity A is greater.

B. Quantity B is greater.

C. The two quantities are equal.

D. The relationship cannot be determined from the information given.

6. If a rectangle with length l and width w is inscribed in a circle with radius 5, which of the following must be true?

A. $w + l = 10$ B. $w + l = 100$ C. $w^2 + l^2 = 25$ D. $w^2 + l^2 = 50$ E. $w^2 + l^2 = 100$

答案及解析

1. The area of a circular region with radius r is 5π.

Quantity A	Quantity B
The area of a circular region with radius $3r$	45π

A. Quantity A is greater.
B. Quantity B is greater.
C. The two quantities are equal.
D. The relationship cannot be determined from the information given.

题干翻译
半径为 r 的圆形区域的面积是 5π。

解题思路
本题需要比较的是半径为 $3r$ 的圆的面积与 45π 的大小关系。
相似图形面积比 = 边长比^2,
因此，半径为 $3r$ 的圆的面积 : 半径为 r 的圆的面积 = $(3r:r)^2 = 9:1 = $ 半径为 $3r$ 的圆的面积 : 5π。
半径为 $3r$ 的圆的面积 = 45π,
故数量 A 与数量 B 相等。

答案 C

2. In the xy-plane, which of the following points is NOT on the circle that has center $(0, 0)$ and radius 5?

A. $(-5, 0)$ B. $(-3, -4)$ C. $(1, -2\sqrt{6})$ D. $(3, 3\sqrt{3})$ E. $(\sqrt{21}, 2)$

题干翻译
在 xy 平面上，下列哪个点不在圆心为 $(0, 0)$、半径为 5 的圆上？

解题思路
半径为 5、圆心为 $(0, 0)$ 的圆的方程为：$x^2 + y^2 = 25$,
在圆上的点需要符合以上方程。
将 D 选项对应的点代入 $x^2 + y^2 = 9 + 27 = 36 \neq 25$，因此 D 选项对应的点不在这个圆上。

答案 D

3. The two circles in the figure shown both have center O. The square is inscribed in the larger circle and circumscribed about the smaller circle. If x is the radius of the larger circle, what is the area of the smaller circle, in terms of x?

A. $\dfrac{3\pi x^2}{4}$ B. $\dfrac{2\pi x^2}{3}$ C. $\dfrac{\pi x^2}{2}$ D. $\dfrac{\pi x^2}{3}$

E. It cannot be determined from the information given.

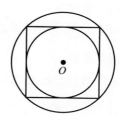

题干翻译

图中的两个圆的圆心为 O。正方形内接在较大的圆内，外切在较小的圆上。如果 x 是大圆的半径，那么小圆的面积是多少？

解题思路

设小圆的半径为 r，小圆的直径 = 外切正方形的边长，因此，正方形的边长 = $2r$。

根据"大圆的直径 = 内接正方形的对角线"可知，大圆的直径 $2x$ = 正方形的对角线 = $2\sqrt{2}r$，

即 $r = \dfrac{\sqrt{2}}{2}x$，

因此，小圆的面积 = $\pi r^2 = \pi\left(\dfrac{\sqrt{2}}{2}x\right)^2 = \dfrac{\pi x^2}{2}$。

答案 C

4. Two concentric circles have diameters 6 and 10. What is the area of the region between the two circles?

A. 64π B. 36π C. 25π D. 16π E. 4π

题干翻译

两个同心圆的直径分别为 6 和 10。两个圆之间的区域面积是多少？

解题思路

根据题干条件可知，两个同心圆的半径分别为 3 和 5。

两个圆之间的环形面积 = 大圆的面积 − 小圆的面积
$$= \pi(5)^2 - \pi(3)^2$$
$$= 16\pi$$

答案 D

5. The area of the shaded portion of the circular region with center O is 16π.

Quantity A	Quantity B
The radius of the circle	5.5

A. Quantity A is greater.

B. Quantity B is greater.

C. The two quantities are equal.

D. The relationship cannot be determined from the information given.

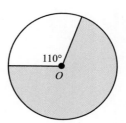

题干翻译

在圆心为 O 的圆形区域内的阴影部分的面积是 16π。

解题思路

本题需要比较圆的半径和 5.5 的大小关系。

设圆的半径为 r，则阴影部分的面积 $= \dfrac{360° - 110°}{360°}\pi r^2 = 16\pi$，

$r = 4.8 < 5.5$。

答案 B

6. If a rectangle with length l and width w is inscribed in a circle with radius 5, which of the following must be true?

A. $w + l = 10$ B. $w + l = 100$ C. $w^2 + l^2 = 25$ D. $w^2 + l^2 = 50$ E. $w^2 + l^2 = 100$

题干翻译

如果一个长为 l、宽为 w 的矩形内接于一个半径为 5 的圆内，下列哪个选项一定正确?

解题思路

根据"圆的直径 = 内接正方形的对角线"可知，

$10 = \sqrt{w^2 + l^2}$，

$w^2 + l^2 = 100$。

答案 E

3.6 立体几何

基本词汇

- rectangular solid 长方体
- cube 立方体
- circular cylinder 圆柱体
- right circular cylinder 正圆柱体
- sphere 球体
- pyramid 锥体
- face 面
- edge 棱
- vertex 顶点
- volume 体积
- surface area 表面积
- lateral surface 侧面

常见的立体图形包括长方体、立方体、圆柱体、球体和锥体。

1. 长方体

长方体和立方体有 6 个面，8 个顶点，12 条边。右图为长 l、宽 w、高 h 的长方体。如果长方体的 6 个面全都是正方形，即 $l = w = h$，那么这个长方体称为正方体。

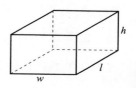

长方体的体积 $= lwh$。

长方体的表面积 $A = 2(wl + wh + lh)$。

2. 立方体

边长为 a 的立方体的体积 $= a^3$。

边长为 a 的立方体的表面积 $= 6a^2$。

3. 圆柱体

圆柱体是由两个完全相等的圆和一个侧面构成。连接两个底面圆心的线段叫作圆柱体的轴。正圆柱体的轴垂直于底面。

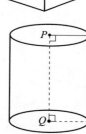

对于底面半径为 r、高为 h 的正圆柱体，它的体积 $V = \pi r^2 h$。

它的表面积 $A = 2\pi r^2 + 2\pi rh$。

4. 球体

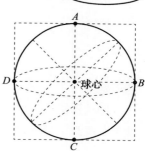

球的体积 $= \dfrac{4}{3}\pi r^3$。

球的表面积 $= 4\pi r^2$。

5. 圆锥

圆锥的体积 = $\frac{1}{3}\pi r^2 h$。

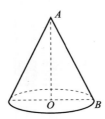

例题 01.

The length of an edge of cube Y is $\frac{2}{3}$ the length of an edge of cube X. What is the ratio of the volume of cube Y to the volume of cube X?

A. 1 to 3　　　B. 1 to 9　　　C. 1 to 27　　　D. 2 to 3　　　E. 8 to 27

题干翻译

立方体 Y 的边长是立方体 X 的边长的 $\frac{2}{3}$，立方体 Y 的体积与立方体 X 的体积之比是多少？

解题思路

体积比 = (边长比)³，

立方体 Y 的体积 : 立方体 X 的体积 = $\left(\frac{2}{3}\right)^3$ = 8 : 27。

答案 E

例题 02.

A cylindrical can, including its two bases, has a total surface area of k square centimeters and a volume of k cubic centimeters. If the height of the can is 8 centimeters, what is the radius, in centimeters, of one of the bases?

A. 2　　　B. $\frac{16}{7}$　　　C. $\frac{8}{3}$　　　D. 4　　　E. 8

题干翻译

一个圆柱形罐子，包括它的两个底座，总表面积为 k 平方厘米，体积为 k 立方厘米。如果罐子的高度是 8 厘米，那么其中一个底座的半径是多少厘米？

解题思路

设圆柱体半径为 r，圆柱体的表面积 $= 2\pi r^2 + 2\pi rh = 2\pi r^2 + 16\pi r = k$。

圆柱体的体积 $= \pi r^2 h = 8\pi r^2 = k$。

即 $2\pi r^2 + 16\pi r = 8\pi r^2$，

$r = \dfrac{8}{3}$。

答案 C

例题 03.

$k > 0$

Quantity A	Quantity B
The volume of a hemisphere (half of a sphere) that has a radius k	The volume of a right circular cylinder that has a height k and whose base has radius k

A. Quantity A is greater.
B. Quantity B is greater.
C. The two quantities are equal.
D. The relationship cannot be determined from the information given.

解题思路

本题需要比较的是半径为 k 的半球的体积与高度为 k、底面圆半径为 k 的圆柱体的体积大小。

半径为 r 的球的体积 $V = \left(\dfrac{4}{3}\right)\pi r^3$，

因此，数量 A 中半径为 k 的半球体积 $= \left(\dfrac{2}{3}\right)\pi r^3$。

底面半径为 r，高为 h 的圆柱的体积为 $= \pi r^2 h$；

数量 B 中半径为 k、高为 k 的圆柱体的体积 $= \pi k^3 > \left(\dfrac{2}{3}\right)\pi k^3$，

即数量 B 更大。

答案 B

例题 04.

The figure shows a right circular cone, where A is the center of the circular base. If the measure of angle ACB is $30°$ and the length of line segment BC is 2, what is the volume of the cone?

(The volume of right circular cone is one-third of the product of the height of the cone and the area of the circular base.)

Give your answer to the <u>nearest whole number</u>.

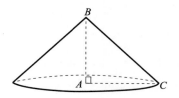

题干翻译

该图显示了一个圆锥体，其中 A 是底部圆的中心。如果 ∠ACB 为 30°，线段 BC 的长度是 2，那么圆锥体的体积是多少？

(圆锥体的体积是圆锥体高度与底面圆的面积乘积的 $\frac{1}{3}$。)

给出最接近的整数答案。

解题思路

$\cos 30° = \dfrac{AC}{BC}$,

$AC = 2 \times \dfrac{\sqrt{3}}{2} = \sqrt{3}$,

$\sin 30° = \dfrac{AB}{BC}$,

$AB = 2 \times \dfrac{1}{2} = 1$,

圆锥的体积 $= \dfrac{1}{3} \pi r^2 \times h = \dfrac{1}{3} \pi (\sqrt{3})^2 \times 1 = \pi \approx 3$。

答案 3

练 习

1. A certain right circular cylinder has volume V. A second right circular cylinder has a radius that is twice the radius of the first cylinder and a height that is twice the height of the first cylinder. What is the volume of the second cylinder in terms of V?

   A. $2V$ B. $4V$ C. $6V$ D. $8V$ E. $16V$

2. A solid wooden peg is in the shape of a right circular cylinder that has diameter 6 millimeters and height 12 millimeters. Approximately what is the surface area, in square centimeters, of the peg?

   A. 2.8   B. 3.4   C. 6.8   D. 9.6   E. 13.6

3. If a rectangular room has length 20 meters, width 10 meters, and height 4 meters, then the volume of the room is how many cubic centimeters?

   A. 80,000
   B. 800,000
   C. 8,000,000
   D. 80,000,000
   E. 800,000,000

答案及解析

1. A certain right circular cylinder has volume V. A second right circular cylinder has a radius that is twice the radius of the first cylinder and a height that is twice the height of the first cylinder. What is the volume of the second cylinder in terms of V?

A. $2V$ B. $4V$ C. $6V$ D. $8V$ E. $16V$

题干翻译

某个圆柱体的体积为 V。第二个圆柱体的半径是第一个圆柱体的半径的 2 倍，且高度是第一个圆柱体高度的 2 倍。那么如何用 V 来表示第二个圆柱体的体积？

解题思路

设第一个圆柱体的半径为 r，则第二个圆柱体的半径为 $2r$，
第一个圆柱体的高度为 h，则第二个圆柱体的半径为 $2h$。
第一个圆柱体的体积 $= \pi r^2 h = V$，
因此，第二个圆柱体的体积 $= \pi (2r)^2 \times 2h = 8\pi r^2 h = 8V$。

答案 D

2. A solid wooden peg is in the shape of a right circular cylinder that has diameter 6 millimeters and height 12 millimeters. Approximately what is the surface area, in square centimeters, of the peg?

A. 2.8   B. 3.4   C. 6.8   D. 9.6   E. 13.6

题干翻译

实心木钉是直径为 6 毫米、高为 12 毫米的圆柱体形状。钉子的表面积大约是多少平方厘米？

解题思路

圆柱体的表面积 $= 2\pi r^2 + 2\pi r h$，
$= 2\pi \times 9 + 6\pi \times 12$ 平方毫米，
≈ 282 平方毫米，
$= 2.82$ 平方厘米。（1 平方厘米 = 100 平方毫米）

答案 A

3. If a rectangular room has length 20 meters, width 10 meters, and height 4 meters, then the volume of the room is how many cubic centimeters?

A. 80,000
B. 800,000
C. 8,000,000
D. 80,000,000
E. 800,000,000

题干翻译

如果一个长方体的房间长 20 米, 宽 10 米, 高 4 米, 那么房间的体积是多少立方厘米?

解题思路

长方体的体积 = 长 × 宽 × 高,
$$= 20m \times 10m \times 4m,$$
$$= 2000cm \times 1000cm \times 400cm,$$
$$= 800000000 cm^3。(1m = 100cm)$$

答案 E

3.7 坐标几何

基本词汇

rectangular coordinate system 平面直角坐标系
xy-coordinate system xy 平面直角坐标系
plane 平面
x-axis x 轴
y-axis y 轴
origin 原点
quadrant 象限
x-coordinate 横坐标
y-coordinate 纵坐标
symmetric about the x-axis 关于 x 轴对称
symmetric about the y-axis 关于 y 轴对称
symmetric about the origin 关于原点对称
Pythagorean theorem 毕达哥拉斯定理
slope 斜率
y-intercept y 轴截距
x-intercept x 轴截距
parallel 平行
perpendicular 垂直
parabola 抛物线
vertex 顶点

概念

在同一个平面上互相垂直且有公共原点的两条数轴构成平面直角坐标系，简称直角坐标系。通常，两条数轴分别置于水平位置与垂直位置，取向右与向上的方向分别为两条数轴的正方向。水平的数轴叫作 x 轴或横轴，垂直的数轴叫作 y 轴或纵轴，x 轴和 y 轴统称为坐标轴，它们的公共原点 O 称为直角坐标系的原点。

考点1 象限

两条坐标轴将平面分成四个区域或者四个象限，分别是第一象限、第二象限、第三象限和第四象限。

第一象限：(＋,＋)，横坐标为正，纵坐标为正。
第二象限：(－,＋)，横坐标为负，纵坐标为正。
第三象限：(－,－)，横坐标为负，纵坐标为负。
第四象限：(＋,－)，横坐标为正，纵坐标为负。
如右图所示。

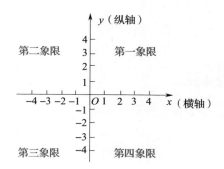

考点 2　平面直角坐标系对称的问题

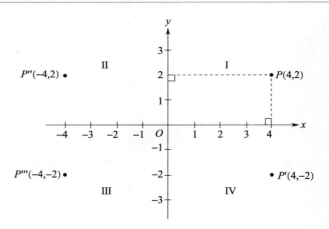

在坐标系中的任意一点 P 都可以用 x 轴和 y 轴的坐标来定义，即用一对坐标值 (x, y) 来定义一个点。x 是横坐标，y 是纵坐标。

在上图中点 P，P'，P''，P''' 有如下关系：

- 点 P 与 P' 关于 x 轴对称；
- 点 P 与 P'' 关于 y 轴对称；
- 点 P 与 P''' 关于原点对称。

两点关于一点对称，该点的横坐标和纵坐标分别是两个端点的横坐标之和的一半和纵坐标之和的一半。例如，A 点 (x_1, y_1) 与 C 点 (x_3, y_3) 关于 B 点 (x_2, y_2) 对称，则

$$\begin{cases} x_2 = \dfrac{x_1 + x_3}{2}, \\ y_2 = \dfrac{y_1 + y_3}{2}. \end{cases}$$

A 点 (x_1, y_1) 关于 x 轴的对称点：$(x_1, -y_1)$。
A 点 (x_1, y_1) 关于 y 轴的对称点：$(-x_1, y_1)$。
A 点 (x_1, y_1) 关于原点的对称点：$(-x_1, -y_1)$。
A 点 (x_1, y_1) 关于直线 $y = x$ 的对称点：(y_1, x_1)。

考点 3　平面直角坐标系上两点间的距离

在平面直角坐标系中，两点之间的距离可以用毕达哥拉斯定理（即勾股定理）求得，例如，A 点 (x_1, y_1) 与 B 点 (x_2, y_2) 间的距离为：$AB = \sqrt{(x_2 - x_1)^2 + (y_2 - y_1)^2}$。

考点 4 直线方程

拥有两个未知数的方程可以在平面直角坐标系中表示出来。在平面直角坐标系中，未知数 x 和未知数 y 表示的方程的图形即所有满足方程的有序数对 (x, y) 在平面中点的集合。

线性方程 $y = kx + b$ 的图形为一条直线，k 被称为直线的斜率，b 代表直线的 y 轴截距。

1. 斜率的计算

如果一条直线经过 A 点 (x_1, y_1) 与 B 点 (x_2, y_2)，$x_1 \neq x_2$，那么这条直线的斜率 $k = \dfrac{y_2 - y_1}{x_2 - x_1}$。

2. 斜率图像

如果直线斜率 >0，直线向右上方倾斜，与 x 轴的正半轴的夹角为锐角，一定经过第一象限和第三象限。

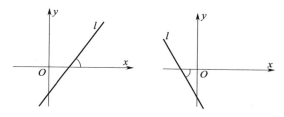

如果直线斜率 <0，直线向右下方倾斜，与 x 轴的正半轴的夹角为钝角，一定经过第二象限和第四象限。
如果直线斜率 =0，则直线和 x 轴平行。
与 y 轴平行的直线没有意义。

3. 斜率与直线平行、垂直的关系

如果两条直线斜率相等（且截距不相等），那么这两条直线互相平行。
如果两条直线互相垂直，那么这两条直线的斜率的乘积为 -1。

4. 截距问题

截距表示的是直线与坐标轴交点的坐标，可以为正，也可以为负。

如果题目要求我们求 x 轴截距，因为 x 轴截距指的是直线与 x 轴的交点，因此将 $y = 0$ 代入直线方程，解得的 x 对应的值就是 x 轴截距。

如果题目要求我们求 y 轴截距，因为 y 轴截距指的是直线与 y 轴的交点，因此将 $x = 0$ 代入直线方程，解得的 y 对应的值就是 y 轴截距。

考点 5 抛物线

一元二次方程 $y = ax^2 + bx + c$（a, b, c 都是实数，且 $a \neq 0$）的图像是一个抛物线。抛物线的 x 轴截距是 $ax^2 + bx + c = 0$ 的根，抛物线的 y 轴截距是常数项 c。

1. 抛物线的开口

如果 $a>0$，那么抛物线开口向上，抛物线的顶点是抛物线的最低点；如果 $a<0$，那么抛物线开口向下，抛物线的顶点是抛物线的最高点。

2. 抛物线的对称轴

对称轴 $x=-\dfrac{b}{2a}$，

如果对称轴 <0，说明对称轴在 y 轴左侧；
如果对称轴 >0，说明对称轴在 y 轴右侧。
在对称轴处，抛物线取最值。

3. 抛物线与 x 轴的交点

- 当 $\triangle = b^2-4ac >0$ 时，抛物线与 x 轴有 2 个交点。
- 当 $\triangle = b^2-4ac =0$ 时，抛物线与 x 轴有 1 个交点。
- 当 $\triangle = b^2-4ac <0$ 时，抛物线与 x 轴没有交点。

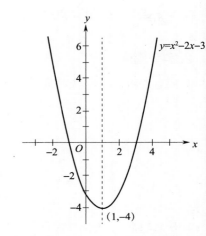

如右图所示，此为抛物线 $y=x^2-2x-3$ 的图像。

在这个图形中，x 轴的截距为 -1 和 3，是通过方程 $x^2-2x-3=0$ 解出来的。点 $(1,-4)$ 是抛物线的顶点，直线 $x=1$ 是抛物线的对称轴。y 轴截距为 -3，在 $x=0$ 时求得。

例题 01.

The four quadrants of the xy-plane are shown. The graph of $y=mx+b$ (not shown), where m and b are positive constants, must pass through which of the quadrants? Indicate all such quadrants.

A. Ⅰ　　B. Ⅱ　　C. Ⅲ　　D. Ⅳ

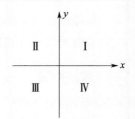

题干翻译

xy 平面的四个象限如右图所示。$y=mx+b$（未示出）的图形，其中 m 和 b 是正常数，一定通过哪些象限？

解题思路

斜率 >0，直线向右上方倾斜，与 x 轴的正半轴的夹角为锐角，一定经过第一象限和第三象限。根据直线方程可知，该直线一定经过点 $(0,b)$，即 y 轴的截距为 b，根据题干条件可知 $b>0$，该直线还经过第二象限。

综上，该直线一定经过第一象限、第二象限和第三象限。

答案　ABC

In the rectangular coordinate system, the line $y = x$ is perpendicular bisector of segment AB (not shown), and the x-axis is the perpendicular bisector of segment BC (not shown). If the coordinates of point A are $(2, 3)$, what are the coordinates of the point C?

A. $(-3, -2)$ B. $(-3, 2)$ C. $(2, -3)$
D. $(3, -2)$ E. $(2, 3)$

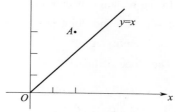

题干翻译

在直角坐标系中，直线 $y=x$ 是线段 AB（未标出）的垂直平分线，x 轴是线段 BC（未标出）的垂直平分线。如果点 A 的坐标是 $(2, 3)$，那么点 C 的坐标是多少？

解题思路

直线 $y=x$ 垂直平分线段 AB，即点 $A(2,3)$ 和点 B 关于直线 $y=x$ 对称，因此点 B 坐标为 $(3, 2)$。x 轴垂直平分线段 BC，即点 B 和点 C 关于 x 轴对称，因此 C 点坐标为 $(3, -2)$。

答案 D

$a \geqslant 1$

In the xy-plane, triangle ABC has vertices A, B, and C. The coordinates of A, B, and C are $(0, 2a)$, $(3a, -a)$, and $(-4a, 0)$, respectively.

Quantity A	Quantity B
AB	AC

A. Quantity A is greater.
B. Quantity B is greater.
C. The two quantities are equal.
D. The relationship cannot be determined from the information given.

题干翻译

在 xy 平面上，三角形 ABC 有顶点 A，B 和 C。A，B 和 C 的坐标分别是 $(0, 2a)$，$(3a, -a)$ 和 $(-4a, 0)$。

解题思路

本题要求比较线段 AB 的长度和线段 AC 的长度的大小。
$AB = \sqrt{(0-3a)^2 + (2a-(-a))^2} = \sqrt{18}a$。
$AC = \sqrt{(0-(-4a))^2 + (2a-0)^2} = \sqrt{20}a$。
因为 $a \geqslant 1$，因此 $\sqrt{18}a < \sqrt{20}a$，故数量 B 更大。

答案 B

例题 04.

In the rectangular coordinate system, point P has coordinates $(-2, 1)$ and point Q has coordinates $(3, 6)$.

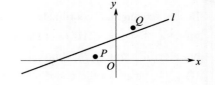

Quantity A	Quantity B
The slope of line l	1

A. Quantity A is greater.
B. Quantity B is greater.
C. The two quantities are equal.
D. The relationship cannot be determined from the information given.

题干翻译

在直角坐标系中，点 P 的坐标为 $(-2, 1)$，点 Q 的坐标为 $(3, 6)$。

解题思路

本题要求比较直线 l 的斜率与 1 的大小关系。

直线 PQ 的斜率 $= \dfrac{6-1}{3-(-2)} = 1$，

直线 l 与 x 轴正半轴的夹角比直线 PQ 与 x 轴正半轴的夹角要小，意味着直线 l 小于直线 PQ 的斜率，因此直线 l 的斜率比 1 小。

故数量 B 更大。

答案 B

例题 05.

In the xy-plane, if line l has equation $3y + 4x = 6$ and line w is perpendicular to line l, what is the slope of line w?

A. $-\dfrac{3}{2}$ B. $-\dfrac{4}{3}$ C. $\dfrac{3}{4}$ D. $\dfrac{4}{3}$ E. $\dfrac{3}{2}$

题干翻译

在 xy 平面上，如果直线 l 的方程为 $3y + 4x = 6$，直线 w 垂直于直线 l，那么直线 w 的斜率是多少？

解题思路

根据题干条件"直线 l 的方程为 $3y + 4x = 6$"可得，

$y = -\dfrac{4}{3}x + 2$，即直线 l 的斜率为 $-\dfrac{4}{3}$。

根据题干条件"直线 w 垂直于直线 l"可知，直线 w 的斜率 × 直线 l 的斜率 $= -1$，

因此直线 w 的斜率 $= \dfrac{3}{4}$。

答案 C

In the *xy*-plane, what is the *y*-intercept of the line determined by the equation $3y - 4x = 12$?

A. -4 B. -3 C. 3 D. 4 E. 12

题干翻译

在 xy 平面上，由方程 $3y - 4x = 12$ 确定的直线的 y 轴截距是多少？

解题思路

求 y 轴截距，即当 $x = 0$ 时，代入直线方程解得的 y 值，

$3y = 12$，

$y = 4$。

答案 D

In the *xy*-plane, the graph of the equation $y = (x+3)(x-2)$ is a parabola.

Quantity A	Quantity B
The *x*-coordinate of the point on the graph for which the *y*-coordinate is a minimum	0

A. Quantity A is greater

B. Quantity B is greater

C. The two quantities are equal

D. The relationship cannot be determined from the information given.

题干翻译

在 xy 平面上，方程 $y = (x+3)(x-2)$ 的图形是抛物线。

解题思路

本题要求比较 y 坐标为最小值的点对应的 x 坐标的值与 0 的大小关系。

抛物线方程 $= (x+3)(x-2) = x^2 + x - 6 = ax^2 + bx + c$，

$a = 1$，$b = 1$，$c = -6$，

$a = 1 > 0$，开口向上，对称轴处取最小值。

对称轴 $x = -\dfrac{b}{2a} = -\dfrac{1}{2} < 0$，

故数量 B 更大。

答案 B

练 习

1. Point P (not shown) is a point on line segment AB. Which of the following could be the coordinates of P?

    A. (2, 2)   B. (3, 1)   C. (4, 3)

    D. (5, 6)   E. (6, 1)

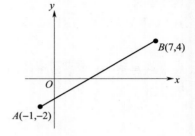

2. In the xy-coordinate system, the distance between the points $(-3, 2)$ and $(5, t)$ is 10.

Quantity A	Quantity B
t	4

A. Quantity A is greater.

B. Quantity B is greater.

C. The two quantities are equal.

D. The relationship cannot be determined from the information given.

3. In the xy-plane, line l has a slope of 2 and contains the point $(1, 7)$. What is the x-intercept of line l?

4. In the xy-plane, the x-intercept and the y-intercept of line j are 3 and -2, respectively, and line k is parallel to j. If the y-intercept of k is 6, what is the x-intercept of k?

答案及解析

1. Point P (not shown) is a point on line segment AB. Which of the following could be the coordinates of P?

    A. (2, 2)   B. (3, 1)   C. (4, 3)

    D. (5, 6)   E. (6, 1)

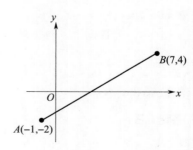

题干翻译

点 P（未标出）是线段 AB 上的点。下列哪一个选项可能是 P 的坐标？

解题思路

设经过 AB 线段的直线方程为：$y = kx + b$。

根据图像可知，线段 AB 过（-1，-2），(7，4) 两点，可得：

$-k + b = -2$，

$7k + b = 4$，

解出 $k = 0.75$，$b = -1.25$。

因此，直线方程为：$y = 0.75x - 1.25$。

点 P 在直线上，需要符合直线方程。

B 选项（3，1）：$0.75 \times 3 - 1.25 = 1$，符合该直线方程。

答案 B

2. In the xy-coordinate system, the distance between the points (-3, 2) and (5, t) is 10.

  Quantity A   Quantity B
    t 4

A. Quantity A is greater.

B. Quantity B is greater.

C. The two quantities are equal.

D. The relationship cannot be determined from the information given.

题干翻译

在 xy 坐标系中，点（-3，2）和点（5，t）之间的距离是 10。

解题思路

本题需要我们比较 t 与 4 的大小关系。

根据两点间的距离公式可知：$10 = \sqrt{(5-(-3))^2 + (t-2)^2} = \sqrt{64 + (t-2)^2}$，

$64 + (t-2)^2 = 100$，

$(t-2)^2 = 36$，

$t - 2 = \pm 6$，

因此，$t = 8(>4)$ 或 $t = -4(<4)$，

故 t 与 4 的大小关系无法确定。

答案 D

3. In the xy-plane, line l has a slope of 2 and contains the point (1, 7). What is the x-intercept of line l?

题干翻译

在 xy 平面上，直线 l 的斜率为 2，包含点 (1，7)。直线 l 的 x 轴截距是多少？

解题思路

设直线 l 的方程为 $y = kx + b$。

根据题干条件"直线 l 的斜率为 2"可知 $k = 2$，即 $y = 2x + b$，

根据题干条件"直线过点 $(1, 7)$"可知 $2 + b = 7$，$b = 5$，

因此，直线方程为：$y = 2x + 5$。

求 x 轴截距，即将 $y = 0$ 代入方程，可得：$2x + 5 = 0$，

$x = -2.5$。

答案 -2.5

4. In the xy-plane, the x-intercept and the y-intercept of line j are 3 and -2, respectively, and line k is parallel to j. If the y-intercept of k is 6, what is the x-intercept of k?

_____

题干翻译

在 xy 平面上，直线 j 的 x 轴截距和 y 轴截距分别为 3 和 -2，直线 k 平行于直线 j。如果直线 k 的 y 轴截距是 6，那么直线 k 的 x 轴截距是多少？

解题思路

根据题干条件"直线 j 的 x 轴截距和 y 轴截距分别为 3 和 -2"可知，直线 j 经过 $(3, 0)$ 和 $(0, -2)$ 两个点，因此可以得到直线 j 的方程 $y = \dfrac{2}{3}x - 2$。

根据题干条件"直线 k 平行于直线 j"可知，直线 k 的斜率为 $\dfrac{2}{3}$。

根据题干条件"直线 k 的 y 轴截距是 6"可得直线 k 的方程为：$y = \dfrac{2}{3}x + 6$。

因此，直线 k 的 x 轴截距，将 $y = 0$ 代入直线 k 的方程，可得 $x = -9$。

答案： -9

第四章 数据分析

数据分析主题包括基本的描述性统计,如平均值、中位数、众数、标准差等;解读表格和图表中的数据,如折条形图、饼图、箱线图、散点图等;初等概率,如互斥事件和独立事件的概率;随机变量和概率分布,包括正态分布;计数方法,如组合、排列和文氏图。这些领域的内容包括高中数学和统计学,但不包括微积分或其他更高层次的数学。

数据分析主要考查 7 个知识点:
(1)描述数据的图表方法
(2)描述数据的数值方法
(3)统计方法
(4)概率
(5)数据分布、随机变量和概率分布
(6)图表题
(7)应用题

4.1 描述数据的图表方法

基本词汇

frequency distribution 频率分布
relative frequency 相对频率
bar graph 条形图
segmented bar graph 分段条形图
circle graph 饼图

pie chart 饼图
sector 扇形
histogram 柱状图
scatter plot 散点图
time plot 时距图

数据可以用许多方法来整合和总结。选择哪种方式与数据本身的性质有关。在代数中变量起着重要的作用。在数据分析中，变量也可以被用来代表我们所分析的个体。比如，人群中的年龄和性别都可以看作变量。变量的分布或者数据分布表明变量的值和在被观察的数据中这个值出现的频率。

1. 频率分布

某一类型或数值的频率是该类型或数值在数据中出现的次数。频率分布是一张表，在这张表中会给出类型或数值，以及它们出现的频率。某一类型或数值的相对频率是该类型或数值出现的频率除以数据的总数。相对频率通常以分数、百分数或小数形式给出。相对频率分布也是一张表，该表中会给出类型或数值以及它们的相对频率。

A survey was taken to find the number of children in each of 25 families. A list of the values collected in the survey follows. 一项调查记录了 25 个家庭中孩子的数量，调查结果如下：

1, 2, 0, 4, 1, 3, 3, 1, 2, 0, 4, 5, 2, 3, 2, 3, 2, 4, 1, 2, 3, 0, 2, 3, 1

以下是这组数据的频率分布和相对频率分布。

Frequency Distribution

Number of Children	Frequency
0	3
1	5
2	7
3	6
4	3
5	1
Total	25

Relative Frequency Distribution

Number of Children	Relative Frequency
0	12%
1	20%
2	28%
3	24%
4	12%
5	4%
Total	100%

> **注意**
> 相对频率的总和为 100%。如果我们使用小数代替百分数，那么总和为 1。

2. 条形图

条形图常被用来表示频率。在条形图中，矩形条用来表示数据种类，每个条形的高度表示该种类的频率或相对频率。所有条形的宽度相同。条形图能够帮助我们比较不同种类出现的频率，我们可以很容易地判断出这一类型的数据是否频繁地出现。

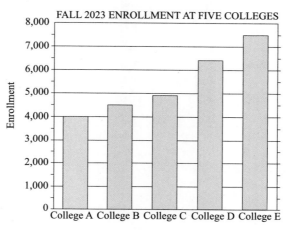

从上图可以看出，2023 年秋季入学人数最多的是 E 大学，最少的是 A 大学，并且根据上图我们可以估算出 D 大学的入学人数大约为 6400 人。

3. 分段条形图

分段条形图被用来展示某个类型是由哪些不同的小组组成。在分段条形图中，每一组都由多个小组组成，每个条形被分成多个条形来表示不同的小组。每个小条形的高度也是根据该小组的频率或相对频率计算出来的。

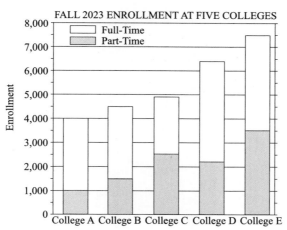

从上图我们可以估算出不同的值。例如，对于 D 大学，总的入学人数大约为 6400 人，非全日制的入学人数大约为 2200 人，全日制的入学人数大约为 6400 − 2200 = 4200 人。条形图还可以用来比较同一种类中的不同组。如下图所示：

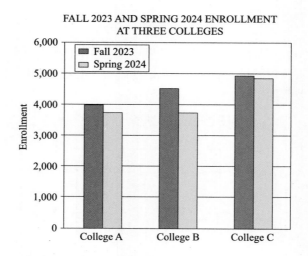

从上图中可知，这三所大学中，2023 年秋季入学人数都比 2024 年春季入学人数多，并且人数减少最多的是大学 B。

虽然条形图经常被用来比较频率，但有时也被用来比较数值数据，比如温度、金钱数量、百分数、高度和重量等。

4．饼图

饼图被用来表述种类较少的数据，它们体现了一个整体是如何被分成各个部分的。饼图中每个类型所占的区域面积与它对应的相对频率成比例。

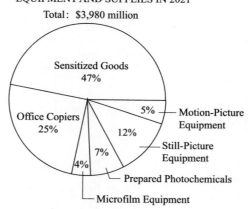

上图呈现了 2021 年美国摄影设备的生产和供应情况，Sensitized Goods 在各个分类中占比最大。饼图中的各个类型所占的区域叫作扇形。因为每个扇形的面积都是根据该类型的相对频率绘制的，因此每个扇形的圆心角与 360° 的比值就等于该种类的相对频率。比如 Prepared Photochemicals 的扇形对应的圆心角 = 7% × 360° = 25.2°。

5. 柱状图

当一组数据较大且包含了许多不同的数值时，我们可以把这组数据分成不同的区间：把全体数据范围分成若干个长度相同的小区间，然后统计落入每个区间的数值的数量。这样，每个区间都有一个频率或相对频率。柱状图通常会呈现各个区间以及它们的频率。

仅有几个值的数值变量也可以用柱状图来表示。每个条形对应的值标注于条形底部的中间。如下图所示。

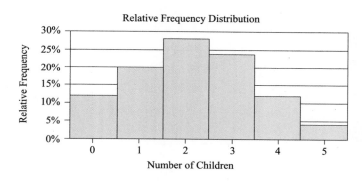

上图就是给出了下表中的相对频率。

Relative Frequency Distribution

Number of Children	Relative Frequency
0	12%
1	20%
2	28%
3	24%
4	12%
5	4%
Total	100%

6. 散点图

此前的例子涉及的数据都是由单一变量产生的，这种类型的数据，即观察一个变量得到的数据被称为单一变量。有时候我们在同一个群体中会研究两个不同的变量，这样的数据被称为二变量数据。我们可能想单独研究这些变量，此前研究单一变量的方法仍然可以使用。而为了体现两个数值变量之间的关系，我们可以用散点图。在散点图中，一个变量的值呈现在直角坐标系的横轴上，另一个变量的值呈现在直角坐标系的纵轴上。对于数据中的每个个体，它的值是一个有序数对，数对中的两个数字各自代表一个变量，这个数将会以点的形式呈现在坐标系中。

我们能从散点图中观察到两个变量之间的关系，或者它们之间关系的一种趋势。

一名自行车教练调查了 50 位自行车爱好者，以检验某项自行车比赛的完成时间与比赛前三个月体能训练量之间的关系。为了衡量训练量，教练开发了一个训练指数，以"单位"来衡量，并基于每名自行车运动员的训练强度。调查所得到的数据以及体现数据趋势的直线呈现在下面的散点图中。

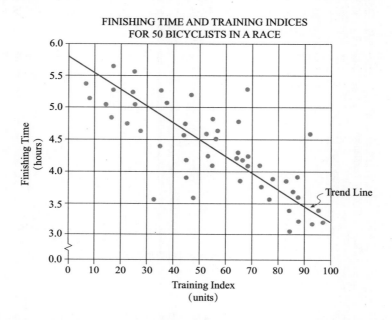

除了图中的趋势线，我们还能够看出散点图中的点偏离这条趋势线的程度。从上图我们能够看出随着训练强度的增加，完成比赛的时间通常会减少。当然，有些记录的数据与这个趋势不符。根据这条趋势线，我们可以做一些预测。比如，当一名自行车爱好者的训练强度为 70 个单位时，他完成比赛大约需要 4 个小时：我们从横轴上 "70" 这个点画一条垂直于横轴的直线与趋势线交于一点，该点的纵坐标非常接近 4 小时。

根据趋势线我们能够做的另一个预测是，自行车爱好者每提高 10 个单位的训练强度能够减少的完成比赛的时间。趋势线的斜率反映的是完成比赛的时间的变化与训练强度变化的比值。趋势线的斜率，我们可以在该直线上任意选取两个点，比如这条趋势线的两个端点：(0, 5.8) 和 (100, 3.2)，趋势线的斜率 $= \frac{3.2 - 5.8}{100 - 0} = -0.026$。这个斜率值体现了当自行车爱好者每提升一个单位的训练强度，他完成比赛所需要的时间将减少 0.026 小时。那么提高十个单位的训练强度他完成比赛所需要的时间将减少 $0.026 \times 10 = 0.26$ 小时。

7. 时距图

时距图用来反映变量如何随时间变化。时距图的横轴为时间，纵轴用来反映测量的变量。在绘制时距图时，我们用线段将两个观察值连接起来，用以强调变量随时间增加或减少。

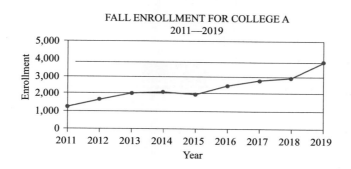

根据上图，我们可以看出在 2018 年到 2019 年之间，秋季入学人数增加得最多。因为 2018 年到 2019 年之间的线段的斜率比其他连续年之间的线段对应的斜率大。

4.2 描述数据的数值方法

基本词汇

- central tendency 集中趋势
- arithmetic mean 平均数
- average 平均数
- median 中位数
- mode 众数
- quartile 四分位数
- range 极差
- interquartile range 四分位数距
- variance 方差
- standard deviation 标准差
- percentile 百分位数
- box plot 箱线图

数据可以用各种统计数据或统计的方法进行描述。这些统计数据通常分为 3 类：数据的集中趋势测度、数据所在位置和整体数据的分布情况。

1. 集中趋势

我们将收集到的数据呈现在数轴上，关于数据的集中趋势，通常涉及平均数、中位数和众数等。

基本概念：

平均数由一组数据中所有数据之和除以这组数据的个数得出。

中位数，又称中值，是按顺序排列的一组数据中居于中间位置的数，我们通常把所有观察值从小到大排序后找出正中间的一个作为中位数。如果观察值为奇数个数，那么这组数据的中位数就是排序后最中间的那个数；如果观察值为偶数个数，中位数取最中间的两个数值的平均数。

平均数很容易受到极值的影响，因为这些值将直接影响到一组数据的总和，从而影响平均数，而中位数不受极值的影响。

注意，等差数列的平均数与中位数相等。

众数是一组数据中出现次数最多的数值，有时众数在一组数中不止一个。

2. 数据所在位置

一组数据从小到大排序后有 3 个最基本的位置：第一个数、最后一个数和中间的那个数。我们通常用 L 代表最小值，G 代表最大值，M 代表中位数。此外，关于数据所在位置的讨论我们还会涉及四分位数和百分位数。中位数、四分位数和百分位数可以将一组从小到大排序好的数据分成若干个大致相同的部分。

3 个四分位数可以把一组数据分成 4 等分，99 个百分位数可以把这组数据分成 100 等分。平均数、中位数、四分位数、百分位数可能不是该组数据中出现的数值。

3 个四分位数从小到大分别为：第一四分位数 Q_1，第二四分位数 Q_2（第二四分位数其实就是中位数），第三四分位数 Q_3。这 3 个数把一组数据分成了 4 个大致相同的部分：第一个部分由最小数 L 到第一四分位数 Q_1 之间的数据构成，第二个部分由第一四分位数 Q_1 到第二四分位数 Q_2 之间的数据构成，第三个部分由第二四分位数 Q_2 到第三四分位数 Q_3 之间的数据构成，第四个部分由第三四分位数 Q_3 到最大数 G 之间的数据构成。因为一组数据中的数据个数不一定能够被 4 整除，所以在不同情况下，我们对于第一四分位数 Q_1 和第三四分位数 Q_3 有着不同的计算规则。但无论什么情况，第二四分位数 Q_2 一定等于中位数 M。

我们最常用的规则：先找第二四分位数 Q_2，也就是中位数 M，中位数 M 将这组数据分成两个相同的部分（较小数的部分和较大数的部分）之后，较小数部分的中位数为第一四分位数 Q_1，较大数部分的中位数为第三四分位数 Q_3。

举例，找出 1，2，3，4，5，6，7，8 这组数据的 3 个四分位数。

首先，我们找到这组数据的第二四分位数 Q_2，也就是这组数据的中位数 $M = \dfrac{4+5}{2} = 4.5$。

中位数 M 将这 8 个数分成（1，2，3，4）和（5，6，7，8）两个等分部分。

第一组数据（1，2，3，4）的中位数即为这 8 个数的第一四分位数 $Q_1 = \dfrac{2+3}{2} = 2.5$，

第二组数据（5，6，7，8）的中位数即为这 8 个数的第三四分位数 $Q_3 = \dfrac{6+7}{2} = 6.5$。

百分位数：将一组数据从小到大排序，刚好处于第 $n\%$ 的位置就成为 n percentile。百分位数适用于数量比较多的数据。99 个百分位数 P_1，P_2，P_3，…，P_{99} 把一组数据分成了 100 等分。其中第一四分位数 Q_1 = 第 25 百分位数 P_{25}，第二四分位数 Q_1 = 中位数 M = 第 50 百分位数 P_{50}，第三四分位数 Q_1 = 第 75 百分位数 P_{75}。由于一组数据的数量不一定能被 100 整除，我们在计算百分位数时也会相应地采用不同的方法。

3. 整体数据的分布情况

我们通常用极差、四分位距和标准差来呈现整体数据的分布情况。

极差指的是一组数据中最大值与最小值之间的差距，极差等于最大值减去最小值。比如，对于数据组：11，8，5，15，21 来说，它的极差 = 21 - 5 = 16。极差反映的是一组数据的最大偏差，但是有时一组数据中会有过大或过小的值。当我们分析一组数据时，会对这些过大或过小的值持怀疑态度，因为这些值可能是由于错误产生的。由于这些异常值在大多数情况下是不会出现的，因此我们分析数据时，通常会忽略这些异常值，而极差很大程度上会受到异常值的影响。

四分位距不受异常值的影响，四分位距为第三四分位数与第一四分位数的差值。

比如，对于我们刚刚谈到的 1，2，3，4，5，6，7，8 这组数据，它的四分位距 = 第三四分位数 - 第一四分位数 = 6.5 - 2.5 = 4。

4. 箱线图

箱线图因形状如箱子而得名，它能显示出一组数据的最大值、最小值、中位数，以及上下四分位数。

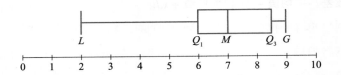

如上图所示，从箱线图中我们可以直接看出这组数据的最小值 L 为 2，第一四分位数 Q_1 为 6，第二四分位数 Q_2 或者中位数 M 为 7，第三四分位数 Q_3 为 8.5，最大值 G 为 9。我们还可以根据这些数据推出，这组数据的极差 $= G - L = 9 - 2 = 7$，这组数据的四分位距 $= Q_3 - Q_1 = 8.5 - 6 = 2.5$。

除此之外，我们还可以从图中看出数据的分布情况。根据四分位数的定义，我们应该知道最小值到第一四分位占了全体数据的 25%；第三四分位到最大值也是占了整体的 25%。虽然它们占的元素数量是一样多的，但是一段比较长、另一段比较短，这说明比较长的那段数据分布比较稀疏，比较短的那一段数据分布比较紧凑。

5. 标准差

与极差和四分位距不同，标准差与一组数据中的每个数据都有联系。标准差所反映的是每个数据与平均数的平均偏离距离。数据离平均数越远，标准差就越大；数据离平均数越近，标准差就越小。

一组数据（假设有 n 个数）标准差的计算：
（1）算出这 n 个数的平均数，
（2）算出每个数与平均数的差值，
（3）对每个差值进行平方，
（4）算出差值的平方的平均数，
（5）算出（4）中得到的平均数的算术平方根。
得到的算术平方根就是这组数据的标准差。

如果一组数据中有 n 个值，x_1，x_2，x_3，…，x_n，平均数为 x，

方差 $= \dfrac{(x_1 - x)^2 + (x_2 - x)^2 + \cdots + (x_n - x)^2}{n}$，

标准差 $= \sqrt{方差} = \sqrt{\dfrac{(x_1 - x)^2 + (x_2 - x)^2 + \cdots + (x_n - x)^2}{n}}$。

关于标准差的相关表达：

x standard deviation(s) above the mean：mean $+ xd$，
比平均值高 x 个标准差：平均值 $+ xd$。

x standard deviation(s) below the mean: mean $- xd$,

比平均值低 x 个标准差：平均值 $- xd$。

x standard deviation(s) of/from the mean: mean $- xd$ 和 mean $+ xd$,

距离平均数 x 个标准差：平均值 $- xd$ 和平均值 $+ xd$。

within x standard deviation(s) of/from the mean: (mean $- xd$) ~ (mean $+ xd$),

在平均值的 x 个标准偏差范围内：（平均值 $- xd$）~（平均值 $+ xd$）。

例如：一组数据平均数为 20，标准差为 5。

1 standard deviation above the mean: mean $+ d = 20 + 5 = 25$。

2 standard deviations below the mean: mean $- 2d = 20 - 10 = 10$。

1 standard deviation of/from the mean: mean $- d$ 和 mean $+ d$，也就是 15 和 25。

within 1 standard deviation of/from the mean: (mean $- d$) ~ (mean $+ d$)，也就是 15 ~ 25。

拓展性质：

（1）一组数据同时增加或减少相同的单位，平均数、中位数、众数发生相同单位的变化，极差、标准差不变。

（2）一组数据同时乘或除以相同的单位，平均数、中位数、众数、极差、标准差发生相同的变化。

例题 01.

The average (arithmetic mean) of x_1, x_2, x_3, x_4 and x_5, is m. Which of the following is the average of $ax_1 + b$, $ax_2 + b$, $ax_3 + b$, $ax_4 + b$, and $ax_5 + b$?

A. am B. $a + b$ C. $m + b$ D. $am + b$ E. $5am + b$

题干翻译

x_1，x_2，x_3，x_4 和 x_5 的平均数为 m。那么 $ax_1 + b$，$ax_2 + b$，$ax_3 + b$，$ax_4 + b$ 和 $ax_5 + b$ 的平均数是多少？

解题思路

根据题干条件可知，$x_1 + x_2 + x_3 + x_4 + x_5 = 5m$，

因此，$ax_1 + b + ax_2 + b + ax_3 + b + ax_4 + b + ax_5 + b = a(x_1 + x_2 + x_3 + x_4 + x_5) + 5b = 5am + 5b$，

$ax_1 + b$，$ax_2 + b$，$ax_3 + b$，$ax_4 + b$ 和 $ax_5 + b$ 的平均数 $= \dfrac{5am + 5b}{5} = am + b$。

答案 D

例题 02.

The sum of 101 consecutive even integers is 20,200.

Quantity A	Quantity B
The average (arithmetic mean) of the 101 integers	The median of the 101 integers

A. Quantity A is greater.
B. Quantity B is greater.
C. The two quantities are equal.
D. The relationship cannot be determined from the information given.

题干翻译

101 个连续偶数的和是 20200。

解题思路

本题需要我们比较这 101 个数的平均数和中位数的大小。

连续偶数是公差为 2 的等差数列，等差数列的平均数 = 中位数。

因此，数量 A = 数量 B。

答案 C

例题 03.

At a certain location, the high temperatures for the past seven days, in degrees Fahrenheit, were 76, 66, 61, 54, 59, x and 70. Which of the following values could be the median of the high temperatures, in degrees Fahrenheit?

Indicate all such values.

A. 59　　　B. 61　　　C. 63　　　D. 64　　　E. 66
F. 67　　　G. 69　　　H. 70

题干翻译

在某个地方，过去 7 天的最高温度，以华氏度计，分别为 76, 66, 61, 54, 59, x 和 70。下列哪个值可能是高温的中位值？

解题思路

7 个数的中位数为第 4 个位置。

先对已有数值排序：

54, 59, 61, 66, 70, 76。

如果 $x \leq 61$，则第 4 个数为 61，B 选项正确，

如果 x 在 61~66 之间，则中位数为 x，CDE 选项正确。

答案 BCDE

The mode of the measurements in a certain list is 47.

     Quantity A   Quantity B

The range of the measurements in the list   47

A. Quantity A is greater.

B. Quantity B is greater.

C. The two quantities are equal.

D. The relationship cannot be determined from the information given.

题干翻译

某个数组中的众数是 47。

解题思路

本题需要我们比较数组的极差和 47 的大小。

极差 = 最大值 − 最小值，而众数指的是出现频率最高的数字，两者无关，因此无法比较。

答案 D

Larry and Tony work for different companies. Larry's salary is the 90th percentile of the salaries in his company, and Tony's salary is the 70th percentile of the salaries in his company.

Which of the following statements individually provides sufficient additional information to conclude that Larry's salary is higher than Tony's salary?

Indicate <u>all</u> such statements.

A. The average (arithmetic mean) salary in Larry's company is higher than the average salary in Tony's company.

B. The median salary in Larry's company is equal to the median salary in Tony's company.

C. The 80th percentile salary in Larry's company is higher than the 70th percentile salary in Tony's company.

题干翻译

Larry 和 Tony 在不同的公司工作。Larry 的工资是他所在公司工资的第 90 百分位，Tony 的工资是他所在公司工资的第 70 百分位。

解题思路

平均数无法计算出 Larry 公司和 Tony 公司某一个具体百分位对应的工资是多少，中位数信息只能提高 Larry 公司和 Tony 公司第 50 百分位对应的工资，所以 AB 都不正确。

C 选项：Larry 公司 80th percentile 的薪水 > Tony 公司 70th percentile 的薪水。根据题干信息，Larry 的薪水是他们公司的 90th percentile，所以 Larry 公司 90th percentile 的薪水 > Larry 公司 80th percentile 的薪水 > Tony 公司 70th percentile 的薪水，从而我们可以判断出 Larry 薪水更高。C 选项正确。

答案 C

例题 06.

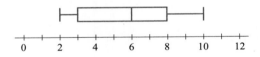

The box plot shown summarizes the total numbers of eggs laid daily by a flock of chickens on a certain farm last June. Based on the information given, which of the following statements must be true?

Indicate all such statements.

A. The flock laid at least 2 eggs a day in all 30 days.
B. The flock laid at least 6 eggs in at least 15 days.
C. The flock laid 10 eggs in at least 1 day.

题干翻译

上图的箱线图总结了去年六月某个农场一群鸡每天下蛋的总数。根据以上给出的信息，下列哪个陈述一定是正确的？

解题思路

根据箱线图可知，2 为最小值，因此 30 天内至少每天下 2 颗蛋，A 选项正确。

6 为中位数，是这 30 天内，从小到大排序后，第 15 个位置和第 16 个位置的平均数。而第 15 个位置对应的数 ≤ 第 15 个和第 16 个位置的平均数，因此至少有 15 天内下了至少 6 颗蛋，B 选项正确。

10 为 30 天内的最大值，因此至少有 1 天下了 10 颗蛋，C 选项正确。

答案 ABC

例题 07.

The standard deviation of x, y, and z is d.

            Quantity A   Quantity B

The standard deviation of $x+1$, $y+1$, and $z+1$ $d+1$

A. Quantity A is greater.

B. Quantity B is greater.

C. The two quantities are equal.

D. The relationship cannot be determined from the information given.

题干翻译

x，y 和 z 的标准差为 d。

解题思路

本题需要我们比较 $x+1$，$y+1$ 和 $z+1$ 这 3 个数的标准差与 $d+1$ 的大小。

一组数据同时增加相同的单位，标准差不变。因此 $x+1$，$y+1$ 和 $z+1$ 这 3 个数的标准差与 x，y 和 z 的标准差相等，为 d，小于 $d+1$，故数量 B 更大。

答案   B

例题 08.

In a data set, the value 79 is $\frac{1}{2}$ standard deviation above the mean and the value 72 is $\frac{2}{3}$ standard deviation below the mean. How many standard deviations above the mean is the value 87? Give your answer as a fraction.

题干翻译

在一个数据集合中，79 比平均值高 $\frac{1}{2}$ 个标准差，72 比平均值低 $\frac{2}{3}$ 个标准差。87 比平均值高多少倍标准差?

解题思路

根据题干条件，可以直接列出方程组:

$$\begin{cases} 79 = m + \frac{1}{2}d, \\ 72 = m - \frac{2}{3}d. \end{cases}$$

解出 $m = 76$，$d = 6$。

因此，87 比平均数 76 多 $\dfrac{87-76}{6} = \dfrac{11}{6}$ 个标准差。

答案 $\dfrac{11}{6}$

练 习

1. What is the average (arithmetic mean) of $\dfrac{a+2}{2}$, $\dfrac{a-1}{2}$, and $\dfrac{2a-1}{2}$?

   A. $4a$ B. $2a$ C. $\dfrac{4a}{3}$ D. $\dfrac{2a}{3}$ E. $\dfrac{a}{2}$

2. $a > 1$

List K：a, $a+1$, $a-1$, $2a$, $-2a$

Quantity A	Quantity B
The average (arithmetic mean) of the five numbers in list K	The median of the five numbers in list K

   A. Quantity A is greater.
   B. Quantity B is greater.
   C. The two quantities are equal.
   D. The relationship cannot be determined from the information given.

3.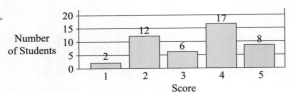

The chart shows the distribution of scores on a quiz in a geometry class.

Quantity A	Quantity B
The median of the quiz scores	The mode of the quiz scores

   A. Quantity A is greater.
   B. Quantity B is greater.
   C. The two quantities are equal.
   D. The relationship cannot be determined from the information given.

4. $40 < r < s < t < 60$

   r, s, and t are even integers.

Quantity A	Quantity B
The range of all possible values of the sum $r + s + t$	36

   A. Quantity A is greater.
   B. Quantity B is greater.
   C. The two quantities are equal.
   D. The relationship cannot be determined from the information given.

5. In the distribution of measurements of the variable x, the mean is 56 and the measurement r lies between the 65th and 70th percentiles. In the distribution of measurements of the variable y, the mean is 56 and the measurement t lies between the 75th and 80th percentiles.

Quantity A	Quantity B
r	t

   A. Quantity A is greater.
   B. Quantity B is greater.
   C. The two quantities are equal.
   D. The relationship cannot be determined from the information given.

6. [Boxplot with scale 20, 25, 30, 35, 40, 45, 50, 55, 60, 65 Seconds]

   A researcher recorded the amount of time, rounded to the nearest second, that it took each of 14 people to complete a certain task. The recorded times are summarized in the boxplot shown. Which of the following statements must be true?

   Indicate all such statements.

   A. At least one person had a recorded time of 25 seconds.
   B. At least one person had a recorded time of 50 seconds.
   C. Each of the 14 people had a recorded time of at most 60 seconds.

7. In the xy-plane, 8 data points lie on the line with equation $y = 40 + 5x$. If the standard deviation of the x-coordinates of the 8 points is 4.6, what is the standard deviation of the y-coordinates of the 8 points?

8. The number 20.0 is 2 standard deviations above the average (arithmetic mean) of the numbers in a certain list, and the number 6.5 is 3 standard deviations below the average of the numbers in the list.

Quantity A	Quantity B
The average of the numbers in the list	14.6

A. Quantity A is greater.

B. Quantity B is greater.

C. The two quantities are equal.

D. The relationship cannot be determined from the information given.

9. Twenty-two measurements are grouped as shown in the frequency distribution.

Interval	Frequency
10–19	3
20–29	6
30–39	10
40–49	1
50–59	2

Quantity A	Quantity B
The range of the 22 measurements	40

A. Quantity A is greater.

B. Quantity B is greater.

C. The two quantities are equal.

D. The relationship cannot be determined from the information given.

答案及解析

1. What is the average (arithmetic mean) of $\frac{a+2}{2}$, $\frac{a-1}{2}$, and $\frac{2a-1}{2}$?

   A. $4a$ B. $2a$ C. $\frac{4a}{3}$ D. $\frac{2a}{3}$ E. $\frac{a}{2}$

题干翻译

$\frac{a+2}{2}$，$\frac{a-1}{2}$和$\frac{2a-1}{2}$的平均数是多少？

解题思路

$$\frac{\left(\frac{a+2}{2}+\frac{a-1}{2}+\frac{2a-1}{2}\right)}{3}=\frac{2a}{3}$$

答案 D

2. $a > 1$

List K: a, $a+1$, $a-1$, $2a$, $-2a$

Quantity A	Quantity B
The average (arithmetic mean) of the five numbers in list K	The median of the five numbers in list K

A. Quantity A is greater.

B. Quantity B is greater.

C. The two quantities are equal.

D. The relationship cannot be determined from the information given.

解题思路

本题让我们比较数组 K 的平均数与中位数的大小。

数组 K 的平均数 $= \dfrac{a+a+1+a-1+2a-2a}{5} = \dfrac{3a}{5}$。

求数组 K 的中位数，结合 $a > 1$，我们先将这 5 个数从小到大排序：$-2a$，$a-1$，a，$a+1$，$2a$。

5 个数的中位数为第 3 个位置对应的数，因此中位数为 a，大于 $\dfrac{3a}{5}$。

故数量 B 更大。

答案 B

3.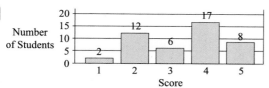

The chart shows the distribution of scores on a quiz in a geometry class.

Quantity A	Quantity B
The median of the quiz scores	The mode of the quiz scores

A. Quantity A is greater.

B. Quantity B is greater.

C. The two quantities are equal.

D. The relationship cannot be determined from the information given.

题干翻译

该图表呈现的是几何课测验的分数分布情况。

解题思路

本题让我们比较测试分数的中位数与众数的大小。

根据图表可知，学生人数 = 2 + 12 + 6 + 17 + 8 = 45，中位数所在位置是第 $\frac{45+1}{2}=23$ 个数字上，因此中位数为 4。

众数指的是出现频率最高的数字，根据图表可知，4 出现频率最高，因此众数为 4。

故数量 A 与数量 B 相等。

答案　C

4. $40 < r < s < t < 60$

r, s, and t are even integers.

    Quantity A   Quantity B

The range of all possible values of the sum $r + s + t$ 36

A. Quantity A is greater.

B. Quantity B is greater.

C. The two quantities are equal.

D. The relationship cannot be determined from the information given.

题干翻译

r, s 和 t 都是偶整数。

解题思路

$r + s + t$ 的最大值 = 58 + 56 + 54 = 168，

$r + s + t$ 的最小值 = 46 + 44 + 42 = 132，

极差 = 最大值 − 最小值 = 168 − 132 = 36，

故数量 A 与数量 B 相等。

答案　C

5. In the distribution of measurements of the variable x, the mean is 56 and the measurement r lies between the 65th and 70th percentiles. In the distribution of measurements of the variable y, the mean is 56 and the measurement t lies between the 75th and 80th percentiles.

Quantity A   Quantity B

  r t

A. Quantity A is greater.

B. Quantity B is greater.

C. The two quantities are equal.

D. The relationship cannot be determined from the information given.

题干翻译

在变量 x 的测量值分布中，平均值为 56，r 位于第 65 和第 70 百分位之间。在变量 y 的测量值分布中，平均值为 56，t 位于第 75 和第 80 百分位之间。

解题思路

百分位对应的数值与平均数无关，题干只给出了两个数组平均数相等这个信息，因此，不能判断 r 和 t 的值。

答案　D

6.

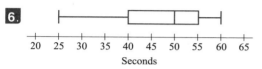

A researcher recorded the amount of time, rounded to the nearest second, that it took each of 14 people to complete a certain task. The recorded times are summarized in the box plot shown. Which of the following statements must be true?

Indicate <u>all</u> such statements.

A. At least one person had a recorded time of 25 seconds.

B. At least one person had a recorded time of 50 seconds.

C. Each of the 14 people had a recorded time of at most 60 seconds.

题干翻译

一名研究人员记录了 14 个人完成某项任务所用的时间，四舍五入到最接近的秒。记录的时间总结在上图所示的箱线图中。下列哪个陈述一定是正确的？

解题思路

根据箱线图可知，第二四分位数 $Q_2 =$ 中位数 $= 50$，第一四分位数 $Q_1 = 40$，第三四分位数 $Q_3 = 55$，最小数 $L = 25$，最大数 $G = 60$。

全体数据中最小值为 25，所以至少有一个数据为 25，A 选项正确。

14 个数据中的中位数 50 是这些数从小到大排序后由第七和第八个数据的平均数得出的，平均数对应的数值在原数据中不一定存在，B 选项不一定成立。

全体数据中最大值为 60，所以每个数据都小于或等于 60，C 选项正确。

答案　AC

7. In the xy-plane, 8 data points lie on the line with equation $y = 40 + 5x$. If the standard deviation of the x-coordinates of the 8 points is 4.6, what is the standard deviation of the y-coordinates of the 8 points?

题干翻译

在 xy 平面上,8 个数据点位于直线方程 $y = 40 + 5x$ 上。如果这 8 个点的 x 坐标的标准差是 4.6,那么这 8 个点的 y 坐标的标准差是多少?

解题思路

将每个数据同时乘 5,标准差变成 $4.6 \times 5 = 23$,

将每个数据同时加 40,标准差不变,还是 23。

答案 23

8. The number 20.0 is 2 standard deviations above the average (arithmetic mean) of the numbers in a certain list, and the number 6.5 is 3 standard deviations below the average of the numbers in the list.

Quantity A	Quantity B
The average of the numbers in the list	14.6

A. Quantity A is greater.

B. Quantity B is greater.

C. The two quantities are equal.

D. The relationship cannot be determined from the information given.

题干翻译

20.0 比某个数组中数字的平均值高 2 个标准差,6.5 比数组中数字的平均值低 3 个标准差。

解题思路

本题需要我们比较数组的平均数和 14.6 的大小。

根据题干条件可列出方程组:

$\begin{cases} m + 2d = 20, \\ m - 3d = 6.5。 \end{cases}$

解得:$m = 14.6$,$d = 2.7$。

数量 A 与数量 B 相等。

答案 C

9. Twenty-two measurements are grouped as shown in the frequency distribution.

Interval	Frequency
10–19	3
20–29	6
30–39	10
40–49	1
50–59	2

   Quantity A   Quantity B
The range of the 22 measurements   40

A. Quantity A is greater.

B. Quantity B is greater.

C. The two quantities are equal.

D. The relationship cannot be determined from the information given.

题干翻译

上面的频率分布表展示的是 22 个测量数据的分布情况。

解题思路

本题需要比较的是这 22 个数据的极差与 40 的大小关系。

极差 = 最大值 - 最小值。

注意，如表所示，有 3 个数据在 10~19 这一范围内，并不意味这 3 个数据中就一定有 10 这个数值。

同理，如表所示，有 2 个数据在 50~59 这一范围内，并不意味这 2 个数据中有 59 这个数值。

因此，极差最大可能的情况是：最大值为 59，最小值为 10，此时极差为 49；极差最小可能的情况是：最大值为 50，最小值为 19，此时极差为 31。

即 31≤数量 A≤49，因此数量 A 与 40 的大小关系不确定。

答案   D

4.3 统计方法

基本词汇

set 集合
element 元素
member 元素
finite set 有限集
infinite set 无限集
empty set 空集
nonempty 非空集
subset 子集
intersection 交集

union 并集
disjointed 互斥
mutually exclusive 互斥
Venn diagram 文氏图/维恩图
universal set 全集
inclusion-exclusion principle 容斥原理
permutation 排列
factorial 阶乘
combination 组合

4.3.1 集合和数组

当一个集合中数据总数较少时，我们可以采用枚举法，即一个一个地数。当一个集合中数据很多，无法采用枚举法，我们就需要用其他方法来计数。

集合是指具有某种特定性质的具体或抽象的对象汇总而成的集体。其中，构成集合的这些对象则称为该集合的元素。有些集合是有限的，即该集合中的元素可以被列举完。有限集可以列举出所有元素并用大括号括起来，例如，偶数集 {0, 2, 4, 6, 8}。非有限集称为无限集，例如，整数集。不包含任何元素的集合叫作空集，空集通常用符号 ∅ 来表示。包含一个或多个元素的集合叫作非空集。如果集合 A 中所有的元素都为集合 B 中的元素，那么集合 A 就是集合 B 的子集。例如，{2, 8} 是 {0, 2, 4, 6, 8} 的子集。空集 ∅ 是所有集合的子集。

数组与有限集类似，数组中的元素可以被列举完，但与有限集不同的是，在数组中，元素是按一定顺序排列的，即将一个数组中的元素重新排列就构成一个新的数组。因此，在数组中，"第一个元素""第二个元素"这样的表达是有意义的。并且，数组中的元素可以重复，且重复是有意义的。例如，数组 1, 2, 3, 2 与数组 1, 2, 2, 3 是不同的数组，每个数组都包含了 4 个元素，并且它们和数组 1, 2, 3 不同，数组 1, 2, 3 只包含 3 个元素。

与数组不同的是，当集合里的元素重复出现时，不另算作一个新的元素，并且元素的顺序没有意义。比如，集合 {1, 2, 3, 2} 和集合 {3, 1, 2} 是同一个集合，都是包含 3 个元素的集合。对于任意一个有限集

S，S 中的元素数量可以用 $|S|$ 表示。比如，如果 $S = \{6.2, -9, \pi, 0.01, 0\}$，那么 $|S| = 5$。$|\varnothing| = 0$。

集合与集合可以一起形成新的集合。集合 S 和集合 T 的交集指的是既属于集合 S 又属于集合 T 的元素组成的集合，记作 $S \cap T$。集合 S 和集合 T 的并集指的是由集合 S 和集合 T 中所有元素组成的集合，记作 $S \cup T$。如果集合 S 和集合 T 没有相同的元素，这两个集合互斥。

考点1　文氏图/维恩图

文氏图，也叫维恩图，能够有效地展现两个或三个集合可能的交集和并集的方式。在文氏图中，我们用圆形区域表示集合。如果集合之间不互斥，那么圆形区域会有重叠；如果集合互斥，那么圆形区域不会有重叠。有时圆形区域会画在一个矩形区域内部，该矩形区域表示全集，所有的集合都是全集的子集。

下面的文氏图展现了集合 A，B，C 以及全集 U。

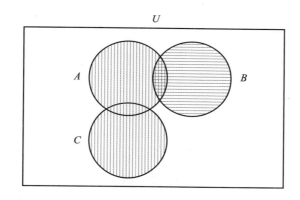

竖条纹区域代表集合 $A \cup C$，横条纹区域代表集合 B。既有竖条纹又有横条纹的区域代表集合 $A \cap B$。集合 B 与集合 C 互斥，记作 $B \cap C = \varnothing$。

在计数时，必须注意没有重复，没有遗漏。为了使重叠部分不被重复计算，人们研究出一种新的计数方法，这种方法的基本思想是：先不考虑重叠的情况，把包含于某内容中的所有对象的数目先计算出来，然后再把计数时重复计算的数目排除出去，使得计算的结果既无遗漏又无重复，这种计数的方法称为容斥原理。这个原理与两个有限集的交集和并集中元素的数量有关：两个集合的并集中的元素等于两个集合中元素数量相加再减去两个集合交集中的元素。

如上图中 $|A \cup B| = |A| + |B| - |A \cap B|$，因为集合 $A \cap B$ 既是集合 A 的子集也是集合 B 的子集，减去 $A \cap B$ 是为了避免重复计算。对于集合 B 与集合 C 的并集，我们有 $|B \cup C| = |B| + |C|$，因为 $B \cap C = \varnothing$。

在 GRE 数学中，两个圆的文氏图是每次考试的必考内容。当题目难度系数变大时，也有可能碰到 3 个圆的文氏图。

1. 2个圆的文氏图

矩形代表全体数据,圆 A 代表符合条件 A 的数据,圆 B 代表符合条件 B 的数据。

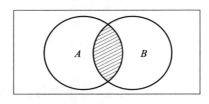

阴影部分代表同时符合 A 和 B 的数据,我们记作 $A \cap B$。

矩形中非圆区域指的是非 A 且非 B 的情况。

我们一般用字母 I 或者 U 代表全体数据,如果题目给出的数据为百分比或者概率,那么全集就是 100% 或 1。

两个圆的文氏图公式:全集 $U = A + B - (A \cap B) +$ 非 A 非 B。

例题 01.

Of the 83 members of a certain club, 48 members enjoy boating and 33 members enjoy fishing. If all except 10 of the members of the club enjoy boating or fishing or both, how many members enjoy boating but not fishing?

A. 15　　　　B. 25　　　　C. 38　　　　D. 40　　　　E. 48

题干翻译

在某个俱乐部的 83 名成员中,48 名成员喜欢划船,33 名成员喜欢钓鱼。如果除了 10 个成员之外,俱乐部的其他成员都喜欢划船或钓鱼,或者两者都喜欢,那么有多少名成员喜欢划船而不喜欢钓鱼?

解题思路

圆 A 表示喜欢划船的人数,圆 B 表示喜欢钓鱼的人数,设喜欢划船且喜欢钓鱼的人数为 x 人。

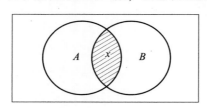

根据"全集公式 $U = A + B - (A \cap B) +$ 非 A 非 B"可得:

$83 = 48 + 33 - x + 10$,

$x = 8$。

因此,喜欢划船而不喜欢钓鱼的人数 = 喜欢划船的人数 - 喜欢划船且喜欢钓鱼的人数,

$= 48 - 8$,

$= 40$。

答案 D

例题 02.

In a group of 45 students, 28 were taking a science class and 35 were taking a math class. If 7 of the students were taking neither a math class nor a science class, how many were taking both a math class and a science class?

A. 11　　　　B. 18　　　　C. 25　　　　D. 28　　　　E. 32

题干翻译

在有 45 名学生的小组中，28 名在上科学课，35 名在上数学课。如果 7 名学生既不上数学课也不上科学课，有多少人同时上数学课和科学课？

解题思路

圆 A 表示上科学课的人数，圆 B 表示上数学课的人数，设既上数学课又上科学课的人数为 x 人。

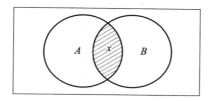

根据"全集公式 $U = A + B - (A \cap B) + 非A非B$"可得：

$45 = 28 + 35 - x + 7$，

$x = 25$。

答案 C

例题 03.

In a group of people, 30 people speak French, 40 speak Spanish, and $\dfrac{1}{2}$ of the people who speak Spanish do not speak French. If each person in the group speaks French, Spanish, or both, which of the following statement are true?

Indicate <u>all</u> such statements.

A. Of the people in the group, 20 speak both French and Spanish.

B. Of the people in the group, 10 speak French but do not speak Spanish.

C. Of the people in the group, $\dfrac{1}{5}$ speak French but do not speak Spanish.

题干翻译

在一组人中，30 人说法语，40 人说西班牙语，$\dfrac{1}{2}$ 的说西班牙语的人不说法语。如果小组中的每个人都说法语、西班牙语或两者都说，下列哪个选项是正确的？

解题思路

圆 A 表示说法语的人数,圆 B 表示说西班牙语的人数,设既说法语又说西班牙语的人数为 x 人。

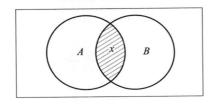

根据题干条件可知,$A=30$,$B=40$,$B-A\cap B=20$,$A\cap B=20$。

根据题干条件"小组中的每个人都说法语、西班牙语或两者都说"可知,非 A 非 $B=0$。

A 选项,这个小组,有 20 个人会说法语和西班牙语。正确。

B 选项,这个小组,有 10 个人会说法语,但不会说西班牙语:只会说法语的人 $=A-A\cap B=30-20=10$,正确。

C 选项,这个小组,$\dfrac{1}{5}$ 的人会说法语,但不会说西班牙语:总人数 $=A+B-(A\cap B)+$ 非 A 非 $B=30+40-20+0=50$,只会说法语的人数 $=A-A\cap B=30-20=10$,只会说法语的人数在总人数中的占比 $=\dfrac{10}{50}=\dfrac{1}{5}$,正确。

答案　ABC

2. 3 个圆的文氏图

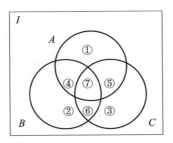

如上图所示,矩形代表全体数据,圆 A 表示符合条件 A 的情况,圆 B 表示符合条件 B 的情况,圆 C 表示符合条件 C 的情况。

由于 3 个圆之间的重合区域比较复杂,所以我们对其进行编号。通过上图我们可以发现,如果 3 个圆直接相加的话,区域①、②、③各出现了 1 次:只有圆 A 里有区域①,只有圆 B 里有区域②,只有圆 C 里有区域③;区域④、⑤、⑥各出现了 2 次:圆 A 和圆 B 中包含区域④,圆 A 和圆 C 中包含区域⑤,圆 B 和圆 C 中包含区域⑥;区域⑦出现了 3 次:圆 A、圆 B 和圆 C 中都出现了区域⑦。

3 个圆的文氏图公式：

全集 $I = A + B + C - (A \cap B) - (A \cap C) - (B \cap C) + (A \cap B \cap C)$ + 非 A 非 B 非 C。

$A \cap B$ 指的是区域④和⑦，$A \cap C$ 指的是区域⑤和⑦，$B \cap C$ 指的是区域⑥和⑦，$A \cap B \cap C$ 指的是区域⑦。

为什么公式里加了一个 $A \cap B \cap C$ 呢？

因为 3 个圆直接相加，区域④⑤⑥出现了 2 次，区域⑦出现了 3 次，所以需要去重。

公式中的 $-(A \cap B) - (A \cap C) - (B \cap C) = -(④+⑦) - (⑤+⑦) - (⑥+⑦)$，

其中区域④⑤⑥重复的那次刚好去除了，而区域⑦被加了 3 次，又减掉了 3 次，即区域⑦被减光了，因此需要把存在的那一次区域⑦给加回来。

所以全集 $I = A + B + C - (④+⑦) - (⑤+⑦) - (⑥+⑦) + (A \cap B \cap C)$ + 非 A 非 B 非 C。

因此，3 个圆的文氏图有 2 个公式：

全集 $I = A + B + C - (A \cap B) - (A \cap C) - (B \cap C) + (A \cap B \cap C)$ + 非 A 非 B 非 C，

全集 $I = A + B + C - (④+⑦) - (⑤+⑦) - (⑥+⑦) + (A \cap B \cap C)$ + 非 A 非 B 非 C。

In a marketing survey for products some people were asked which of the products, if any, they use. Of the people surveyed, a total of 400 use A, a total of 400 use B, a total of 450 use C, a total of 200 use A and B simultaneously, a total of 175 use B and C simultaneously, a total of 200 use A and C simultaneously, a total of 75 use A, B, and C simultaneously, and a total of 200 use none of the products. How many people were surveyed?

A. 950　　　　B. 975　　　　C. 1000　　　　D. 1025　　　　E. 1050

题干翻译

在一项产品的市场调查中，有人被问及他们使用哪种产品（如果有的话）。在接受调查的人中，有 400 人使用 A，有 400 人使用 B，有 450 人使用 C，有 200 人同时使用 A 和 B，有 175 人同时使用 B 和 C，有 200 人同时使用 A 和 C，有 75 人同时使用 A、B 和 C，有 200 人不使用任何产品。有多少人接受了调查？

解题思路

全集 $I = A + B + C - (A \cap B) - (A \cap C) - (B \cap C) + (A \cap B \cap C)$ + 非 A 非 B 非 C，

$= 400 + 400 + 450 - 200 - 200 - 175 + 75 + 200$，

$= 950$。

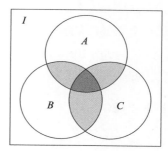

答案　A

例题 02.

Of the 300 subjects who participated in an experiment using virtual-reality therapy to reduce their fear of heights, 40 percent experienced sweaty palms, 30 percent experienced vomiting, and 75 percent experienced dizziness. If all of the subjects experienced at least one of these effects and 35 percent of the subjects experienced exactly two of these effects, how many of the subjects experienced only one of these effects?

A. 105　　　　B. 125　　　　C. 130　　　　D. 180　　　　E. 195

题干翻译

在参与使用虚拟－现实疗法减少恐高症实验的 300 名受试者中，40% 的人经历了手心出汗，30% 的人经历了呕吐，75% 的人经历了头晕。如果所有的受试者都经历了这些反应中的至少一种，而 35% 的受试者只经历了其中的两种反应，那么有多少受试者只经历了其中的一种反应？

解题思路

圆 A 代表手心出汗的人数：$A = 40\% \times 300 = 120$，

圆 B 代表呕吐的人数：$B = 30\% \times 300 = 90$，

圆 C 代表眩晕的人数：$C = 75\% \times 300 = 225$，

只经历两种反应对应的是④⑤⑥区域，

只经历一种反应对应的是①②③区域。

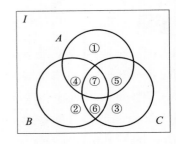

根据题干条件"所有人都经历至少一种反应"可知，

非 A 非 B 非 $C = 0$。

根据题干条件"35% 的受试者只经历了其中的两种反应"可知，④ + ⑤ + ⑥ = 35% × 300 = 105 人。

本题需要我们求出只经历一种反应的人数，即① + ② + ③。

$I = A + B + C - (④ + ⑦) - (⑤ + ⑦) - (⑥ + ⑦) + (A \cap B \cap C) +$ 非 A 非 B 非 C，

$I = A + B + C - (④ + ⑤ + ⑥) - 2 \times ⑦$，

$300 = 120 + 90 + 225 - 105 - 2 \times ⑦$，

⑦ = 15。

根据全集公式和题干条件① + ② + ③ + ④ + ⑤ + ⑥ + ⑦ = I，可得：

① + ② + ③ + 105 + 15 = 300，

① + ② + ③ = 180。

答案　180

例题 03.

In a certain class, 10 students can play the piano, 14 students can play the violin, 11 students can play the flute. If 3 students can play exactly three instruments, 20 students can play exactly one instrument, how many students can play exactly two instruments?

A. 3   B. 6   C. 9   D. 12   E. 18

题干翻译

在一个班级里,10 个学生会弹钢琴,14 个学生会拉小提琴,11 个学生会吹笛子。如果 3 个学生能演奏三种乐器,20 个学生只能演奏一种乐器,那么有多少学生只能演奏两种乐器?

解题思路

圆 A 代表会弹钢琴的人数:$A = 10$,

圆 B 代表会拉小提琴的人数:$B = 14$,

圆 C 代表会吹笛子的人数:$C = 11$。

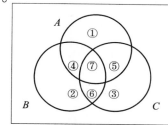

三种乐器都会的人数⑦ = 3,只会一种乐器的人数① + ② + ③ = 20 人,题目需要我们算出只会两种乐器的人数,即④ + ⑤ + ⑥。

$A + B + C = (① + ② + ③) + 2 \times (④ + ⑤ + ⑥) + 3 \times ⑦$,

$10 + 14 + 11 = 20 + 2 \times (④ + ⑤ + ⑥) + 3 \times 3$,

④ + ⑤ + ⑥ = 3。

答案 A

3. 双重标准分组

双重标准分组,指的是题目中提到了两种分类依据。按照两种分类依据,可以把所有数据分为两种。双重标准分组的题目,我们通常画 4×4 的表格来求解。

例题 01.

In a group of 106 people, 58 are teachers and 28 are at least 50 years old. If 16 are teachers who are at least 50 years old, how many in the group are not teachers and are less than 50 years old?

A. 12   B. 36   C. 42   D. 48   E. 70

题干翻译

在106人的小组中，58人是教师，28人至少50岁。如果16人是至少50岁的教师，那么这个小组中有多少人不是教师并且不到50岁？

解题思路

根据题干信息可知，对这个小组有两套分类标准：是不是教师，以及年龄是不是50岁以上。因此，本题应该采用双重标准分组来解决。

	Teacher	Not teacher	
≥50	16		28
<50		?	106 − 28 = 78
	58	106 − 58 = 48	106

根据题干信息可知，不是教师的人 = 106 − 58 = 48 人，

不是教师且年龄≥50的人 = 28 − 16 = 12 人，

因此，不是教师且<50岁的人 = 48 − 12 = 36 人。

答案 B

例题 02.

Of the 72 fish in a certain aquarium, $\frac{1}{3}$ are brown and the remaining $\frac{2}{3}$ are white. Among the 72 fish, $\frac{1}{2}$ have orange fins and $\frac{1}{2}$ have black fins. If $\frac{2}{3}$ of the white fish have orange fins, how many of the brown fish have black fins?

A. 4　　　B. 16　　　C. 20　　　D. 24　　　E. 32

题干翻译

在某个水族馆的72条鱼中，$\frac{1}{3}$是棕色的，剩下的$\frac{2}{3}$是白色的。在72条鱼中，$\frac{1}{2}$的鱼有橙色的鳍，$\frac{1}{2}$的鱼有黑色的鳍。如果$\frac{2}{3}$的白鱼有橙色的鳍，那么有多少棕色的鱼有黑色的鳍？

解题思路

根据题干信息可知，对这个小组有2套分类标准：鱼的颜色，以及鱼鳍的颜色。因此，本题应该采用双重标准分组来解决。

	Brown	White	
Orange fins		$\frac{2}{3} \times 48 = 32$	$\frac{1}{2} \times 72 = 36$
Black fins	?		$\frac{1}{2} \times 72 = 36$
	$\frac{1}{3} \times 72 = 24$	$\frac{2}{3} \times 72 = 48$	72

根据题干信息可知，有橙色的鳍的白鱼数量 = $\frac{2}{3} \times 48 = 32$，

因此，有橙色的鳍的棕鱼数量 = 36 − 32 = 4，

有黑色的鳍的棕鱼数量 = 棕色的鱼的总数 − 有橙色的鳍的棕鱼数量 = 24 − 4 = 20。

答案 C

At a medical clinic, there are exactly 20 medical personnel — 8 doctors and 12 nurses. Of these medical personnel, 9 are men, and 3 of the 9 men are nurses. If a doctor is chosen at random, what is the probability that the doctor chosen will be a woman?

A. $\frac{1}{10}$ B. $\frac{3}{20}$ C. $\frac{1}{4}$ D. $\frac{3}{8}$ E. $\frac{11}{20}$

题干翻译

在一家医疗诊所，有 20 名医务人员——8 名医生和 12 名护士。在这些医务人员中，9 名是男性，9 名男性中有 3 名是护士。如果随机选择一个医生，选择的医生是女性的概率有多大？

解题思路

根据题干信息可知，对这个小组有 2 套分类标准：职务是医生还是护士，以及性别。因此，本题应该采用双重标准分组来解决。

	Women	Men	
Doctor	?		8
Nurse		3	12
	20 − 9 = 11	9	20

根据题干信息可知，男医生人数 = 9 − 3 = 6，

因此，女医生人数 = 8 − 6 = 2，

从 8 名医生中选出女医生的概率 = $\frac{2}{8} = \frac{1}{4}$。

答案 C

4.3.2 排列组合

1. 组合

如果我们从 A，B，C，D 这 4 个字母中选出 3 个字母，不需要对这 3 个字母进行排序，那么一共有多少

种情况呢？下面我们列举出了所有的情况：

ABC，ABD，ACD，BCD。

在不考虑排序的情况下，从 4 个字母中选出 3 个字母一共有 4 种不同的情况。

我们如果从 m 个对象中选出 n 个对象（$n \leq m$），但不对选出的 n 个对象进行排序，这种单纯的抽选叫作组合。我们一般用 C（Combination）表示组合，组合公式如下：

$$C_m^n = \frac{m!}{n!(m-n)!}。$$

在具体的运算中，组合的公式可以进行化简。比如：$C_{10}^2 = \frac{10!}{2!(10-2)!} = \frac{10!}{2!8!} = \frac{10 \times 9}{2 \times 1} = 45$。

2. 排列

从 m 个对象中选出 n 个对象，并且对这 n 个对象进行排序。

比如，从 A，B，C，D 这 4 个字母中选出 3 个字母，并要求对这 3 个字母进行排序，有多少种可能呢？

这就是一个典型的排列问题，因为谁是第一个字母，谁是第二个字母，谁是第三个字母代表了不同的结果。

排列一般用字母 P（Permutation）或者字母 A（Arrangement）来表示，排列公式如下：

$$P_m^n \text{ 或 } A_m^n = \frac{m!}{(m-n)!}。$$

在具体的运算中，组合的公式可以进行化简。

比如：$P_{10}^2 = \frac{10!}{(10-2)!} = \frac{10!}{8!} = 10 \times 9 = 90$。

3. 加法原理和乘法原理

我们在处理排列组合的问题时，会碰到需要分类讨论的情况。那么在汇总分类讨论的结果时，我们应该进行加减运算还是乘除运算呢？

如果分类讨论的情况，属于整个事情的一个分类，那么进行加减运算；

如果分类讨论的情况，属于整个事情的一个步骤，那么进行乘除运算。

例如：

①小明要选择餐厅吃饭，他可以吃西餐或者中餐。如果西餐厅有 5 种选择，中餐厅有 3 种选择，那么小明可以选择的餐厅有 $5 + 3 = 8$ 种。因为无论是选择中餐厅还是选择西餐厅，都可以跟小明选择餐厅这件事画上等号。

②小明要去餐馆点餐，这顿饭包括主菜和甜品。如果菜单中主菜有 5 种选择，甜品有 3 种选择，这顿饭必须由 1 个主菜和 1 个甜品构成，那么这顿饭小明有 $5 \times 3 = 15$ 种选择。因为单独点主菜或者单独点甜品，都不能算完成了这顿饭的点餐要求，必须选择 1 个主菜和 1 个甜品这顿餐才算点完。

在GRE中，排列组合通常考以下几种情况。
① 直接代公式，
② 分组抽选，
③ 依次讨论，
④ 正难则反，
⑤ 捆绑问题，
⑥ 重复元素问题。

考点 1　直接代公式

认真读题，判断题目描述的是组合问题还是排列问题，然后将数据代入对应的公式进行计算即可。

例题 01.

A certain club has 10 members. What is the ratio of the number of 5-member committees that can be formed from the member of the club to the number of 4-member committees that can be formed from the member of the club?

题干翻译

某俱乐部有10名成员。由俱乐部成员组成的5人委员会的数量与由俱乐部成员组成的4人委员会的数量之比是多少？

解题思路

从10个人中选出5个人或者从10个人中选出4个人，不涉及顺序，因此是一个组合问题。

10个人中选出5个人的委员会数量 $= C_{10}^{5}$，

10个人中选出4个人的委员会数量 $= C_{10}^{4}$，

$$\frac{C_{10}^{5}}{C_{10}^{4}} = \frac{\frac{10!}{5!5!}}{\frac{10!}{4!6!}} = \frac{10!}{5!5!} \times \frac{4!6!}{10!} = \frac{6}{5}$$。

答案　$\dfrac{6}{5}$

例题 02.

At a college graduation ceremony, there will be 5 speakers seated in a row of 5 seats on a stage. How many possible seating arrangements for the 5 speakers are there?

题干翻译

在一个大学毕业典礼上，将有 5 位演讲者坐在舞台上的一排 5 个座位上。这 5 位发言者有多少种可能的座位安排？

解题思路

由于 ABCDE 和 EDCBA 是完全不同的安排，说明本题中的座位安排是一个排列问题。

因此，这 5 个人的座位排序 = $P_5^5 = 5! = 5 \times 4 \times 3 \times 2 \times 1 = 120$。

答案　120

例题 03

A knockoff website requires users to create a password using letters from the word MAGOSH. If each password must have at least 4 letters and no repeated letters area allowed, how many different passwords are possible?

题干翻译

某山寨网站要求用户使用单词 MAGOSH 中的字母创建密码。如果每个密码必须至少有 4 个字母，并且字母不允许重复使用，那么有多少种不同的密码？

解题思路

注意，密码问题和号码问题都是要考虑顺序的，属于排列问题。

如果密码由 4 个字母构成，那么从单词 MAGOSH 的 6 个字母中选出 4 个字母进行排列，共 $P_6^4 = 360$ 种可能。

如果密码由 5 个字母构成，那么从单词 MAGOSH 的 6 个字母中选出 5 个字母进行排列，共 $P_6^5 = 720$ 种可能。

如果密码由 6 个字母构成，那么从单词 MAGOSH 的 6 个字母中选出 6 个字母进行排列，共 $P_6^6 = 720$ 种可能。

因此，密码共有 $360 + 720 + 720 = 1800$ 种可能。

答案　1800

考点 2　分组抽选

"分组抽选"和"直接抽选"有什么区别呢？

我们来看一下具体的例子：

情况 1：一个班里有 10 个学生，从中抽出 4 个学生构成一个学生委员会，有多少种可能？

有 C_{10}^4 种可能。

情况2：一个班里有10个学生，包括5个男生5个女生，从中抽出由2个男生和2个女生构成的4人学生委员会，有多少种可能？

这种情况下，答案就不是C_{10}^4了。

从5个女生中抽2个女生，有C_5^2种可能；从5个男生中抽2个男生，有C_5^2种可能。2个男生+2个女生一起才能构成本题要求的4人学生委员会，因此中间用乘号相连，即有$C_5^2 \times C_5^2$种可能。

 分组抽选指的是将数据分成不同的组，从各个组中分别抽取题目要求的对应元素，一起构成一个整体。

Set K consists of 9 positive integers, 5 of which are prime numbers. How many subsets of K consist of 3 integers such that 2 integers are prime numbers and 1 integer is not a prime number?

题干翻译

集合K由9个正整数组成，其中5个整数是质数。K有多少个子集由3个整数组成，这3个整数中2个整数是质数，1个整数不是质数？

解题思路

根据题干条件，这9个正整数中5个是质数，那么余下的4个就不是质数。

选出的3个整数的要求：2个是质数，1个不是质数，3个数字都选出来才算完成，所以中间用乘号相连。

共$C_5^2 \times C_4^1 = 10 \times 4 = 40$。

答案 40

Two companies C_1 and C_2 are participating in a fund-raising activity along with 8 other companies. Of the 10 companies, a group of 4 companies will be chosen to receive an award. Of all the possible choices of groups of 4 companies that will receive an award, how many choices include both companies C_1 and C_2? (Enter your answer as an integer or a decimal in the answer box. Backspace to erase.)

题干翻译

两家公司C_1和C_2与其他8家公司一起参加了一项筹资活动。在这10家公司中，将选出4家公司获奖。在将获奖的4家公司的所有可能选择中，有多少种选择包括公司C_1和C_2？

解题思路

根据题干条件，这 4 家获奖的公司必须包含：C_1、C_2 和其他两家公司。

选出公司 C_1 的可能性 = C_1^1（公司 C_1 只有 1 家，且必须选上），

选出公司 C_2 的可能性 = C_1^1（公司 C_2 只有 1 家，且必须选上），

从剩下的非 C_1 和 C_2 的 8 家公司里选出 2 家公司的可能性 = C_8^2，

4 家公司都选出来才算完成，所以中间用乘号相连，

共 $C_1^1 \times C_1^1 \times C_8^2 = 1 \times 1 \times 28$ 种可能。

答案 28

In a university, a certain committee consists of 6 faculty members, 4 administrators and 3 students. A subcommittee of 5 members will be selected from the committee. Professor Smith, who is one of the 6 faculty members, and Ms. Wilson, who is one of the 4 administrators must be on the subcommittee. The other 3 subcommittee members will be selected at random from the rest of the committee, how many different 5-member subcommittees can be selected?

A. 45　　　　B. 72　　　　C. 165　　　　D. 720　　　　E. 1,287

题干翻译

在一所大学里，一个委员会由 6 名教师、4 名行政人员和 3 名学生组成。现在从委员会中选出一个由 5 名成员组成的小组委员会。6 名教员之一的 Smith 教授和 4 名行政人员之一的 Wilson 女士必须是小组委员会的成员。其他 3 名小组委员会成员将从委员会的剩余成员中随机选出，可以选出多少种不同的 5 人小组委员会？

解题思路

根据题干条件，这 5 人小组委员会必须包含：Smith、Wilson 和其他 3 名成员。

选出 Smith 的可能性 = C_1^1（教员 Smith 只有 1 个人，且必须选上），

选出 Wilson 的可能性 = C_1^1（行政人员 Wilson 只有 1 个人，且必须选上），

从剩下的 11 名成员中选出 3 名成员的可能性 = C_{11}^3，

5 人都选出来才算完成，所以中间用乘号相连。

共 $C_1^1 \times C_1^1 \times C_{11}^3 = 1 \times 1 \times 165$ 种可能。

答案 C

考点3 依次讨论

"依次讨论"适用于两种情况：数位问题和依次抽选问题。

1. 数位问题

数字、号码是有顺序的，属于排列问题。比如，从0~9中同时抽取3个数字构成一个号码，结果为P_{10}^3。但如果要求从0~9中抽取3个数字构成一个号码，该号码百位不能为0，且个位必须是奇数，结果就不是P_{10}^3了。或者从0~9中抽取3个数字构成一个号码，号码各个位置的数字可以重复，结果也不是P_{10}^3。

无论排列问题还是组合问题，元素都是同时抽选的，既然是同时抽选，而不是抽完再放回，那么同一个元素是不可能被选2次的。

因此，考试中的数位问题，即问几位数有几种可能的问题或密码问题，如果涉及限制条件或者元素允许重复，我们不应使用排列公式P_m^n，比较推荐的方法是：直接罗列各个数位分别有几种可能性，然后再将每个位置对应的可能性相乘，得到的结果就是总的可能性。注意，如果涉及限制条件，我们要优先处理有限制条件的位置；如果多个位置存在限制，优先处理限制最多的位置，再处理限制次数多的位置，以此类推。

例题 01.

A certain code used for communication consists of sequences of characters, where each character is a dot or a dash. If all possible sequences of 2 characters, 3 characters, and 4 characters are in the code, which of the following statements are true about the sequences in the code?

Indicate <u>all</u> such statements.

A. There are 4 different sequences of 2 characters each.
B. There are 8 different sequences of 3 characters each.
C. There are 16 different sequences of 4 characters each.

题干翻译

用于通信的特定代码由字符序列组成，其中每个字符是一个点或破折号。如果代码中包含所有可能的2个字符、3个字符和4个字符的序列，那么关于代码中的序列，下列哪个陈述是正确的？

解题思路

密码问题属于数位问题，每个位置分别讨论可能性，再相乘，得到总的可能性。
根据题干条件"每个字符是一个点或破折号"可知，每个位置有2种可能的选择。
2个字符的代码的可能性 = 2×2 = 4，A选项正确。
3个字符的代码的可能性 = 2×2×2 = 8，B选项正确。
4个字符的代码的可能性 = 2×2×2×2 = 16，C选项正确。

答案 ABC

例题 02.

$S = \{1, 2, 3\}$
$T = \{1, 2, 3, 4\}$

Quantity A	Quantity B
The number of 4-digit positive integers that can be formed using only the digits in set S	The number of 3-digit positive integers that can be formed using only the digits in set T

A. Quantity A is greater.

B. Quantity B is greater.

C. The two quantities are equal.

D. The relationship cannot be determined from the information given.

解题思路

本题要求我们比较的是数字只用集合 S 中元素形成的 4 位数的可能性与数字只用集合 T 中元素形成的 3 位数的可能性的大小关系。

数位问题采用依次讨论的方法。

只用集合 S 中元素形成的 4 位数的可能性：4 位数，4 个位置，每个位置有 3 种可能的选择，因此，可能性 $= 3 \times 3 \times 3 \times 3 = 81$。

只用集合 T 中元素形成的 3 位数的可能性：3 位数，3 个位置，每个位置有 4 种可能的选择，因此，可能性 $= 4 \times 4 \times 4 = 64$（$<81$）。

故数量 A 更大。

答案　A

例题 03.

If m is an even integer greater than 2000 that is composed from 1, 2, 3, and 4 with no repetition, how many possibilities of m?

题干翻译

如果 m 是一个大于 2000 的偶整数，由 1，2，3，4 组成，数位对应的数字没有重复，m 有多少种可能性？

解题思路

本题对于数字 m 有多个位置限制，我们来分别看一看。

四位数，各个位置的数字来自 1，2，3，4。要求是偶数，因此末位数字只有 2 和 4 两种选择。

数字 $m > 2000$，首位数字有 2，3，4 三种选择。

比对限制条件，我们可以看出末位限制最多，先排末位，再排首位，再排其他数位。注意题干还有"各个数位上的数字不能重复"这个要求，因此这个四位数的可能性＝末位的可能性×首位的可能性×第二个数位的可能性×第三个数位的可能性＝2×2×2×1=8。

答案 8

2. 依次抽选问题

依次抽选和普通的抽选有什么区别呢？我们来看一下具体的例子。

情况1：一个班里有10个学生，包括5个男生5个女生，从中抽出由2个男生和2个女生构成的4人学生委员会，有多少种可能？

从5个女生中抽2个女生，有C_5^2种可能；从5个男生中抽2个男生，有C_5^2种可能。2个男生＋2个女生一起才能构成本题要求的4人学生委员会，因此中间用乘号相连，即有$C_5^2 \times C_5^2$种可能。

情况2：一个班里有10个学生，包括5个男生5个女生，从中抽出4个学生，一次只能抽一个学生，前两次抽女生，后两次抽男生，有多少种可能性？

此时答案就不是$C_5^2 \times C_5^2$了。

情况1中不涉及顺序问题，但情况2中前两次是女生后两次是男生，涉及顺序问题。

因此，未来题干中出现了 one at a time（一次只能抽一个）或者 each time（每一次）这两个标志词，属于依次抽选问题。依次抽选不适合直接使用排列P_m^n或组合C_m^n的公式，比较推荐的方法是：直接罗列每次对应的可能性，然后将可能性相乘，得到的结果就是总的可能性。

以刚刚的情况2为例：

第一次抽出的是女生的可能性：5种。

第二次抽出的是女生的可能性：4种。

第三次抽出的是男生的可能性：5种。

第四次抽出的是男生的可能性：4种。

所以总的可能性是：5×4×5×4=400种。

A bag contains 13 blue pens and 3 red pens. Two pens are to be randomly selected from the bag, one at a time and without replacement. What is the probability that both of the pens selected will be blue? (Enter your answer as an integer or a decimal in the answer box. Backspace to erase.)

题干翻译

一个包里有 13 支蓝笔和 3 支红笔。从袋子中随机选择两支笔,一次选一支,不得更换。被选出的两支笔都是蓝色的概率是多少?

解题思路

根据题干关键词 one at a time 可知,这是一个依次讨论问题,每个位置单独讨论,最后相乘即可。

第一次抽选出蓝笔的概率 = $\frac{13}{16}$,

第二次抽选出蓝笔的概率 = $\frac{12}{15}$,

因此,被选出的两支笔都是蓝笔的概率 = $\frac{13}{16} \times \frac{12}{15} = 0.65$(题干要求填入整数或小数)。

答案 0.65

例题 02

A gardener is going to plant 2 red rosebushes and 2 white rosebushes. If the gardener is to select each of the bushes at random, one at a time, and plant them in a row, what is the probability that the 2 rosebushes in the middle of the row will be the red rosebushes?

A. $\frac{1}{12}$ B. $\frac{1}{6}$ C. $\frac{1}{5}$ D. $\frac{1}{3}$ E. $\frac{1}{2}$

题干翻译

一个园丁打算种 2 盆红花和 2 盆白花。如果园丁一次随机选择种一盆花,种成一排,那么中间的 2 盆是红花的概率是多少?

解题思路

根据题干关键词 one at a time 可知,这是一个依次讨论问题。

共 4 盆花,中间 2 盆是红花,意味着两旁的花是白花。

第 1 盆种白花的概率:4 盆花中 2 盆白花随机选一盆白花的概率为 $\frac{2}{4}$,

第 2 盆种红花的概率:剩下 3 盆花中 2 盆红花随机选一盆红花的概率为 $\frac{2}{3}$,

第 3 盆种红花的概率:剩下 2 盆花中 1 盆红花随机选一盆红花的概率为 $\frac{1}{2}$,

第 4 盆种白花的概率:剩下 1 盆白花中选一盆白花的概率为 $\frac{1}{1}$,

因此总概率 = $\frac{2}{4} \times \frac{2}{3} \times \frac{1}{2} \times \frac{1}{1} = \frac{1}{6}$。

答案 B

考点4 正难则反

正难则反指的是正面解决一个问题非常复杂，我们可以反向思考。

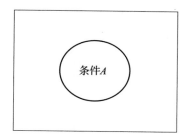

题目要求我们算出条件A发生的可能性，如果直接算条件A的可能性特别复杂，我们可以先算出总的可能性，然后再算出与条件A完全相反的可能性，这样条件A的可能性 = 总的可能性 – 不符合条件A的可能性。

例如，有8本杂志，4本运动杂志，4本时尚杂志，从中要选出4本杂志，且这4本杂志中至少有1本运动杂志，有多少种可能性？

4本杂志中至少有1本运动杂志，可能是：1本运动杂志3本时尚杂志，2本运动杂志2本时尚杂志，3本运动杂志1本时尚杂志，4本运动杂志。如果正着算的话，需要分上述4种情况讨论再汇总：

$C_4^1 \times C_4^3 + C_4^2 \times C_4^2 + C_4^3 \times C_4^1 + C_4^4 = 16 + 36 + 16 + 1 = 69$ 种。

如果采用正难则反，我们需要算出总的可能性，从8本杂志中选出4本杂志的总的可能性 = C_8^4 = 70 种。

4本杂志中至少有1本运动杂志的反向是4本都不是运动杂志的可能性，即4本都是时尚杂志，4本都是时尚杂志的可能性 = $C_4^4 = 1$。

因此，4本杂志中至少有1本运动杂志的可能性 = 总的可能性 – 4本都不是运动杂志的可能性 = 70 – 1 = 69。

通过比对我们发现，正面讨论的结果和正难则反计算的结果相同，但正难则反的步骤更少，解题效率更高。未来在题干中如果出现了 not/cannot（不能）或者 at least one（至少1个）这两个标志词之一时，我们一般用正难则反的方法解题。

at least one（至少1个）的反向是"1个都没有"，not/cannot（不能）的反向是"偏偏可以"。

There are 8 books on a shelf, of which 2 are paperbacks and 6 are hardbacks. How many possible selections of 4 books from this shelf include at least one paperback?

A. 40   B. 45   C. 50   D. 55   E. 60

题干翻译

书架上有 8 本书，其中 2 本是平装本，6 本是精装本。从这个书架上选出 4 本书，这 4 本书中至少有 1 本平装书的可能性是多少？

解题思路

题干出现了 at least one 这一关键词，应该利用正难则反来解题。

4 本书中至少有 1 本平装书的可能性 = 总的可能性 − 4 本书都不是平装书的可能性。

总的可能性 = $C_8^4 = 70$，

4 本书都不是平装书的可能性（4 本书都是精装书的可能性）= $C_6^4 = 15$，

因此，4 本书中至少有 1 本平装书的可能性 = 70 − 15 = 55。

答案 D

例题 02.

$R = \{1, 3, 4, 6, 9\}$
$T = \{2, 4, 5, 7, 9\}$

Pairs of integers are to be selected from the sets shown above: the first integer from set R and the second integer from set T. Of all possible selections, how many pairs include at least one 4?

A. 8   B. 9   C. 10   D. 16   E. 25

题干翻译

从上面的集合中选择整数对：从集合 R 中选出第一个整数并且从集合 T 中选出第二个整数。在所有可能的选择中，有多少对至少包括一个 4？

解题思路

题干中出现了 at least one 这一关键词，应该利用正难则反来解题。

数对中至少包含 1 个 4 的可能性 = 总的可能性 − 没有一个数是 4 的可能性
$$= 5 \times 5 - 4 \times 4 = 9。$$

答案 B

例题 03.

Boxes A, B, and C each contain colored pencils. Box A contains 7 pencils, 3 of which are red. Box B contains 6 pencils, 4 of which are red. Box C contains 7 pencils, 4 of which are red. One pencil is to be selected at random from each box. What is the probability that at least 1 of the 3 pencils selected will be red?

A. $\dfrac{2}{7}$ B. $\dfrac{11}{20}$ C. $\dfrac{45}{49}$ D. $\dfrac{8}{9}$ E. $\dfrac{17}{18}$

题干翻译

盒子 A，B 和 C 中各装有彩色铅笔。盒子 A 里有 7 支铅笔，其中 3 支是红色的。盒子 B 里有 6 支铅笔，其中 4 支是红色的。盒子 C 里有 7 支铅笔，其中 4 支是红色的。从每个盒子里随机抽取一支铅笔，选出的 3 支铅笔中至少有 1 支是红色的概率是多少？

解题思路

根据题干条件"盒子 A 中 7 支铅笔，3 支是红色的"可知，盒子 A 中 4 支铅笔不是红色的。
根据题干条件"盒子 B 中 6 支铅笔，4 支是红色的"可知，盒子 B 中 2 支铅笔不是红色的。
根据题干条件"盒子 C 中 7 支铅笔，4 支是红色的"可知，盒子 A 中 3 支铅笔不是红色的。
题干出现了 at least 1 这一关键词，应该利用正难则反来解题。
至少有 1 支红色铅笔的概率 = 总的概率 − 都不是红色铅笔的概率

$$= 1 - \dfrac{4}{7} \times \dfrac{2}{6} \times \dfrac{3}{7}$$

$$= 1 - \dfrac{4}{49}$$

$$= \dfrac{45}{49}$$

答案 C

考点 5　捆绑问题

如果题干中出现了"A 和 B 必须挨在一起"或者"A 和 B 不能挨在一起（有人把 A 和 B 隔开）"，可以用捆绑法解题。

"必须挨着"的英文表达：next to each other，cannot be separated。

标志词 1：A 和 B 必须挨在一起

比如，A，B，C，D，E 是 5 个朋友，排成一排拍照，因为 A 和 B 关系非常要好，要求必须挨着拍照，那么一共有多少种排法？

对于 A 和 B 必须挨在一起的情况，我们会先把 A 和 B 打包，看作 1 个整体，然后对总的 4 个元素进行排序，即 P_4^4。注意，A 和 B 虽然要求挨在一起，但是还存在 A 和 B 之间的内部排序，所以我们还要对 A 和 B 之间进行内部排序，即 P_2^2。因此总的排法 = $P_4^4 \times P_2^2$。

捆绑法的具体步骤：
①将必须挨在一起的元素打包看成一个整体，
②对打包元素进行内部排序，
③打包后，对全体元素进行外部排序，
④全体元素外部排序 × 打包元素内部排序。

标志词 2：A 和 B 不能挨在一起

A，B，C，D，E 是 5 个朋友，排成一排拍照，因为 A 和 B 关系决裂，不能挨着拍照，那么一共有多少种排法？

出现了关键词"不能"，我们可以考虑正难则反。

A 和 B 不能挨在一起的可能性 = 总的可能性 − A 和 B 必须挨在一起的可能性，
$$= P_5^5 - P_4^4 \times P_2^2。$$

A science book, a philosophy book, a geometry book, and a trigonometry book are to be arranged next to each other on a shelf. There are n possible arrangements of the 4 books such that the geometry book and the trigonometry book are next to each other.

Quantity A　　　　Quantity B
　　n　　　　　　　10

A. Quantity A is greater.
B. Quantity B is greater.
C. The two quantities are equal.
D. The relationship cannot be determined from the information given.

题干翻译
1 本科学书、1 本哲学书、1 本几何书和 1 本三角学书要挨着放在书架上。这 4 本书中几何书和三角学书挨着放的情况有 n 种可能的排列方式。

解题思路
本题需要我们比较 n 与 10 的大小关系。
题干中出现了关键词 next to（挨着），考虑捆绑法。
要求几何书和三角学书挨着（捆绑）的排法：$n = P_3^3 \times P_2^2 = 3 \times 2 \times 1 \times 2 \times 1 = 12$（>10）。

答案　A

Darrell has a collection of 40 DVDs, each of which contains one movie. There are 17 comedy movies, 14 fantasy movies, and 9 historical movies, where each historical movie takes place in a separate time period. The movies will be ordered on a shelf from left to right so that the movies of each type comedy, fantasy and historical are in a single group of consecutive movies. In addition, the historical movies will be ordered from the earliest time period to the latest time period. How many possible orderings of the movies are there?

A. $(17!)(14!)(3!)$
B. $(17!)(14!)(9!)$
C. $(17!)(14!)(9!)(3!)$
D. $\dfrac{40!}{(17!)(14!)(9!)}$
E. $\dfrac{(40!)(3!)}{(17!)(14!)}$

题干翻译

Darrell 收藏了 40 张 DVD，每张都包含一部电影。喜剧电影 17 部，奇幻电影 14 部，历史电影 9 部，每部历史电影都发生在一个独立的时间段。电影将从左到右排列在架子上，这样每种类型的喜剧、奇幻和历史电影都在一组连续的电影中（即喜剧在喜剧组，奇幻在奇幻组，历史在历史组）。此外，历史电影将从最早的时间段到最晚的时间段排序。这些电影共有多少种可能的顺序？

解题思路

优先用捆绑法，
① 将各个类型的电影捆绑，捆绑成 3 组，对捆绑后的 3 组进行排序：3!
② 对每个类型内部进行排序：
   comedy movies 组内排序：17!，
   fantasy movies 组内排序：14!。
   根据题干条件"historical movies 从最早的时间段到最晚的时间段排序"，可知顺序是固定，只有 1 种情况。
③ 总可能性 = (3!)(17!)(14!)。

答案 A

考点6 重复元素问题

什么是重复元素问题呢？

如果我们对 A，B，C，D，E 这 5 个字母进行排序，有 P_5^5 种可能。但如果要求我们对 A，A，B，B，C 这 5 个字母进行排序，就不是 P_5^5 种可能了。因为 2 个 A 彼此之间交换位置，2 个 B 之间彼此交换位置，得到的结果并没有不同。如果我们套用P_5^5来算，会重复计算相同的情况，因此我们需要去除重复的情况。既然2个A彼此之间交换位置，得到的结果并没有不同，那么我们需要去除2个A之间的内部排序P_2^2；2个B彼此之间交换位置，得到的结果并没有不同，那么我们需要去除2个B之间的内部排序P_2^2。注意，2个A之间的内部排序并不等于5个字母的排序，同理，2个B之间的内部排序并不等于5个字母的排序，即它们不能和5个字母排序画上等号，它们只是5个字母排序中的一部分，所以我们应该进行乘除运算，而不是加减运算。因此A，A，B，B，C这5个字母进行排序的可能性 = $\dfrac{P_5^5}{P_2^2 \times P_2^2}$。

重复元素问题公式 = $\dfrac{P_n^n}{P_a^a \times P_b^b}$（其中 n 表示总的元素个数，a 表示一个元素重复了 a 次，b 表示一个元素重复了 b 次）。

例题 01.

In how many distinguishable ways can the 7 letters in the world MINIMUM be arranged, if all the letters are used each time?

A. 7　　　　B. 42　　　　C. 420　　　　D. 840　　　　E. 5040

题干翻译

对单词 MINIMUM 的 7 个字母进行重新排序，每个字母都被使用，有多少种不同的排列方式？

解题思路

在单词 MINIMUM 中，M 重复了 3 次，I 重复了 2 次，

因此可能性 = $\dfrac{P_n^n}{P_a^a \times P_b^b} = \dfrac{P_7^7}{P_3^3 \times P_2^2} = 420$。

答案　C

例题 02.

A certain type of code is a list of 4 identical asterisks and 2 identical dots in any order. For example, **. *. * is one such code.

Quantity A	Quantity B
The number of possible codes of this type	15

A. Quantity A is greater.

B. Quantity B is greater.

C. The two quantities are equal.

D. The relationship cannot be determined from the information given.

题干翻译

某种类型的代码是由 4 个相同的星号和 2 个相同的点以任意顺序组成。比如 ∗∗.∗.∗ 就是其中一种代码。

解题思路

本题要求比较代码的可能性与 15 的大小关系。

根据题干可知,"∗"重复了 4 次,"."重复了 2 次,

因此代码可能性 $= \dfrac{P_n^n}{P_a^a \times P_b^b} = \dfrac{P_6^6}{P_4^4 \times P_2^2} = 15$,

故数量 A 与数量 B 相等。

答案 C

Quantity A	Quantity B
The number of different positive 4-digit integers in which two of the digits are equal to 5 and the other two digits are equal to 7	6

A. Quantity A is greater.

B. Quantity B is greater.

C. The two quantities are equal.

D. The relationship cannot be determined from the information given.

解题思路

本题要求比较由 2 个 5 和 2 个 7 形成的四位数的可能性与 6 的大小关系。

根据题干可知,5 重复了 2 次,7 重复了 2 次,

因此代码可能性 $= \dfrac{P_n^n}{P_a^a \times P_b^b} = \dfrac{P_4^4}{P_2^2 \times P_2^2} = 6$。

故数量 A 与数量 B 相等。

答案 C

4.4 概率

基本词汇

probability 概率　　　exclusive events 互斥事件　　　independent events 独立事件

概念

概率是反映随机事件出现的可能性大小。随机事件是指在相同条件下，可能出现也可能不出现的事件。例如，从一批有正品和次品的商品中，随意抽取一件，"抽得的是正品"就是一个随机事件。

事件概率的取值范围为 0 到 1 之间的数字，反映的是实验中事件发生的可能性。概率对应的值越大，说明事件发生的可能性越大。

"可能性"和"概率"的区别：

"可能性"指的是符合条件的事情发生有多少种可能，它的结果是一个整数。

"概率"指的是在 $\dfrac{\text{符合条件的事情的可能性}}{\text{没有限定条件的情况下的可能性}}$，它的结果是一个 0 到 1 之间的数字。

普通的概率问题和事件概率问题的区别：

普通的概率问题主要是和排列组合结合起来考查。在排列组合中，碰到的题目会要求计算概率的情况，用我们上面讲的在 $\dfrac{\text{符合条件的事情的可能性}}{\text{没有限定条件的情况下的可能性}}$ 计算即可。

事件概率问题属于概率论的内容。事件与事件之间存在联系，比如互斥关系、独立关系、包含关系等，在 GRE 数学中，常考的是互斥关系和独立关系。

考点1　互斥关系

互斥事件：事件 A 和事件 B 不会同时发生。

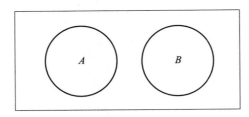

如上图所示：圆 A 表示事件 A 的概率，圆 B 表示事件 B 的概率，两个圆之间不存在交集。

关于互斥事件，我们需要掌握 2 个公式：

互斥事件 A 和 B 同时发生的概率 $P(A \cap B) = 0$，

互斥事件 A 和 B 至少有一个发生的概率 $P(A \cup B) = P(A) + P(B)$。

互斥事件的标志词有 exclusive，cannot both occur，disjoint。

考点 2　独立关系

独立事件：事件 A 和事件 B 相互独立，不会互相影响，可能同时发生，也可能 A 发生了 B 没发生。如下图所示：圆 A 表示事件 A 发生的概率，圆 B 表示事件 B 发生的概率。

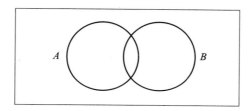

关于独立事件，我们需要掌握 2 个公式：

独立事件 A 和 B 同时发生的概率 $P(A \cap B) = P(A) \times P(B)$，

独立事件 A 和 B 至少有一个发生的概率 $P(A \cup B) = P(A) + P(B) - P(A) \times P(B)$。

独立事件的标志词有 independent。

In a probability experiment R, S, and T are events such that $P(S) = P(T) = x$. $P(R) = kx$, and $P(R \text{ or } S) < P(S \text{ or } T)$, where k and x are positive numbers. Events R and S are mutually exclusive, and events S and T are independent.

Quantity A　　　　　　　　Quantity B

k　　　　　　　　　　　　$1 - x$

A. Quantity A is greater.

B. Quantity B is greater.

C. The two quantities are equal.

D. The relationship cannot be determined from the information given.

题干翻译

在概率实验中，R，S 和 T 的概率 $P(S) = P(T) = x$。$P(R) = kx$，$P(R \text{ or } S) < P(S \text{ or } T)$，其中 k 和 x 为正数。事件 R 和 S 是互斥的，事件 S 和 T 是独立的。

解题思路

本题需要比较 k 与 $1-x$ 的大小。

根据题干条件"事件 R 和 S 互斥"可知，$P(R \text{ or } S) = P(R) + P(S) = kx + x$；

根据题干条件"事件 S 和 T 独立"可知，$P(S \text{ or } T) = P(S) + P(T) - P(S) \times P(T) = 2x - x^2$；

根据题干条件"$P(R \text{ or } S) < P(S \text{ or } T)$"可得：

$kx + x < 2x - x^2$，

$x^2 + kx - x < 0$，

$x(x + k - 1) < 0$。

不等式两边同时除以正数 k 可得：

$x + k - 1 < 0$，

$k < 1 - x$。

故数量 B 更大。

答案 B

To wake up in the morning, Doug sets two alarm clocks that operate independently of each other, in case one alarm clock fails to ring. If the probability that the first clock will ring is 0.95 and the probability that the second clock will ring is 0.90, what is the probability that neither alarm clock will ring?

题干翻译

为了早上醒来，Doug 设置了两个独立运行的闹钟，以防其中一个闹钟没响。如果第一个闹钟响的概率是 0.95，第二个闹钟响的概率是 0.90，那么两个闹钟都不响的概率是多少？

解题思路

根据题干标志词 independently 可知，两个事件为独立事件。

因此，第一个闹钟和第二个闹钟都不响的概率为：

$P(\text{第一个不响} \cap \text{第二个不响}) = P(\text{第一个不响}) \times P(\text{第二个不响})$，

$= (1 - 0.95)(1 - 0.90)$，

$= 0.005$。

答案 0.005

The probability that event R will occur is $\dfrac{4}{5}$, and the probability that event S will occur is $\dfrac{4}{7}$. R and S are independent events.

Quantity A	Quantity B
The probability that both events R and S will occur	The probability that either event R or event S, but not both, will occur

A. Quantity A is greater.
B. Quantity B is greater.
C. The two quantities are equal.
D. The relationship cannot be determined from the information given.

题干翻译

事件 R 发生的概率是 $\frac{4}{5}$，事件 S 发生的概率是 $\frac{4}{7}$。R 和 S 是独立的事件。

解题思路

本题需要比较的是事件 R 和事件 S 都发生的概率与只有事件 R 或只有事件 S 发生的概率的大小。根据题干标志词 independent 可知，R 和 S 为独立事件。

事件 R 和事件 S 都发生的概率 $P(R \cap S) = P(R) \times P(S) = \frac{4}{5} \times \frac{4}{7} = \frac{16}{35}$。

只有事件 R 或只有事件 S 发生且两者不会同时发生的概率

$= P(R \cup S) - P(R \cap S)$，

$= P(R) + P(S) - P(R) \times P(S) - P(R) \times P(S)$，

$= \frac{4}{5} + \frac{4}{7} - \frac{4}{5} \times \frac{4}{7} - \frac{4}{5} \times \frac{4}{7}$，

$= \frac{16}{35}$。

故数量 A = 数量 B。

答案 C

考点3 未知事件

题目没有给出明确的事件关系，那这个事件就属于未知事件。

未知事件 A 和 B 至少有一个发生的概率 $P(A \cup B) = P(A) + P(B) - P(A \cap B)$。

如果题干中没有给出 $P(A \cap B)$ 或 $P(A \cup B)$ 的具体信息，我们只能考虑是否得出 $P(A \cap B)$ 或 $P(A \cup B)$ 的可能范围。

If the probability that event A will occur is 0.5 and the probability that event B will occur is 0.4, which of the following could be the probability that both events will occur?

Indicate all such values.

A. 0　　　B. 0.1　　　C. 0.4　　　D. 0.5　　　E. 0.9

题干翻译

如果事件 A 发生的概率是 0.5，事件 B 发生的概率是 0.4，下列哪一项可能是两个事件都发生的概率？

解题思路

由于题干未给出事件 A 与事件 B 的关系，因此属于未知事件。

当事件 A 与事件 B 互斥时，事件 A 和事件 B 同时发生的概率最小：$P(A \cap B) = 0$。

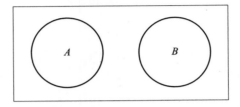

当事件 B 完全被事件 A 包含时，事件 A 和事件 B 同时发生的概率最大：$P(A \cap B) = P(B) = 0.4$。

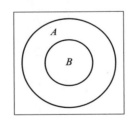

因此 $0 \leq P(A \cap B) \leq 0.4$。

答案　ABC

If the probability that event A will occur is 0.5 and the probability that event B will occur is 0.6, which of the following could be the probability that both events will occur?

Indicate all such values.

A. 0　　　B. 0.1　　　C. 0.4　　　D. 0.5　　　E. 0.9

题干翻译

如果事件 A 发生的概率是 0.5，事件 B 发生的概率是 0.6，下列哪一项可能是两个事件都发生的概率？

> **解题思路**
>
> 由于题干未给出事件 A 与事件 B 的关系，因此属于未知事件。
>
> 注意，本题中事件 A 发生的概率与事件 B 发生的概率相加在一起超过 1，所以事件 A 与事件 B 不可能互斥。
>
> $P(A \cup B) \leq 1$,
>
> $P(A) + P(B) - P(A \cap B) \leq 1$,
>
> $0.5 + 0.6 - P(A \cap B) \leq 1$,
>
> $P(A \cap B) \geq 0.1$。
>
> 当事件 B 完全被事件 A 包含时，事件 A 和事件 B 同时发生的概率最大：$P(A \cap B) = P(A) = 0.5$，因此 $0.1 \leq P(A \cap B) \leq 0.5$。
>
> **答案** BCD

练 习

1. One piece of candy is to be selected at random from each of 4 different boxes of assorted candies. For each of the boxes, the probability that the candy selected will be a chocolate candy is $\frac{1}{10}$, and the 4 selections are to be made independently of each other.

Quantity A	Quantity B
The probability that none of the 4 pieces of candy selected will be a chocolate candy	$\frac{4}{5}$

   A. Quantity A is greater.
   B. Quantity B is greater.
   C. The two quantities are equal.
   D. The relationship cannot be determined from the information given.

2. The probability that event M occurs is 0.3, and the probability that event S occurs is 0.6. If M and S are independent, then the probability that neither of them occurs is

   A. 0.10　　B. 0.18　　C. 0.28　　D. 0.60　　E. 0.82

3. If the probability that event R will occur is 0.75 and the probability that event M will occur is 0.58, which of the following is equal to the maximum possible probability that both events will occur?

A. 0.58 B. 0.75 C. 0.58 + 0.75

D. $\dfrac{0.58 + 0.75}{2}$ E. $0.58 + 0.75 - (0.58)(0.75)$

4. The probability that event R will occur is 0.35, and the probability that events R and T will both occur is p. What is the least possible value of p?

A. 0.00 B. 0.35 C. 0.50 D. 0.65 E. 0.70

5. The probability that both events A and B will occur is $\dfrac{1}{2}$. Which of the following values could be the probability that event A will occur?
Indicate all such values.

A. 0 B. $\dfrac{1}{4}$ C. $\dfrac{1}{2}$ D. $\dfrac{3}{4}$ E. 1

6. The probability that event R occurs is 0.25, and the probability that event T occurs is 0.20. If these events cannot both occur, what is the probability that neither of them occurs?

A. 0.95 B. 0.55 C. 0.50 D. 0.45 E. 0.05

答案及解析

1. One piece of candy is to be selected at random from each of 4 different boxes of assorted candies. For each of the boxes, the probability that the candy selected will be a chocolate candy is $\dfrac{1}{10}$, and the 4 selections are to be made independently of each other.

Quantity A	Quantity B
The probability that none of the 4 pieces of candy selected will be a chocolate candy	$\dfrac{4}{5}$

A. Quantity A is greater.
B. Quantity B is greater.
C. The two quantities are equal.

D. The relationship cannot be determined from the information given.

题干翻译

有4盒不同的什锦糖果，从每盒中随机抽取一块糖果。对于每个盒子，选择的糖果是巧克力糖果的概率是 $\frac{1}{10}$，并且这4个选择是相互独立的。

解题思路

本题需要比较的是4块糖都不是巧克力的概率与 $\frac{4}{5}$ 的大小关系。

根据题干标志词 independently 可知，"选择4块糖果"是独立事件。

4块糖都不是巧克力的概率 $= \left(1 - \frac{1}{10}\right)^4 = 0.6561 \left(< \frac{4}{5} / 0.8\right)$，

故数量 B 更大。

答案 B

2. The probability that event M occurs is 0.3, and the probability that event S occurs is 0.6. If M and S are independent, then the probability that neither of them occurs is

A. 0.10　　　　B. 0.18　　　　C. 0.28　　　　D. 0.60　　　　E. 0.82

题干翻译

事件 M 发生的概率是0.3，事件 S 发生的概率是0.6。如果事件 M 和 S 是独立的，那么它们都不发生的概率是多少。

解题思路

根据题干标志词 independent 可知，M 和 S 为独立事件。

因此 M 和 S 都不发生的概率 $= P(M\text{不发生}) \times P(S\text{不发生}) = (1-0.3) \times (1-0.6) = 0.28$。

答案 C

3. If the probability that event R will occur is 0.75 and the probability that event M will occur is 0.58, which of the following is equal to the maximum possible probability that both events will occur?

A. 0.58　　　　B. 0.75　　　　C. 0.58 + 0.75

D. $\frac{0.58 + 0.75}{2}$　　　　E. $0.58 + 0.75 - (0.58)(0.75)$

题干翻译

如果事件 R 发生的概率是0.75，事件 M 发生的概率是0.58，下列哪一项等于两个事件都发生的最大可能概率？

解题思路

题干没说事件类型，属于未知事件。

未知事件两个都发生概率最大的情况是在概率小的事件完全被概率大的事件包含的情况下，此时两个事件都发生的概率等于概率小的事件的整体，即 0.58。

答案 A

4. The probability that event R will occur is 0.35, and the probability that events R and T will both occur is p. What is the least possible value of p?

A. 0.00 B. 0.35 C. 0.50 D. 0.65 E. 0.70

题干翻译

事件 R 发生的概率是 0.35，事件 R 和 T 同时发生的概率是 p。p 的最小可能值是多少？

解题思路

题干没说事件类型，属于未知事件。

当事件 T 和事件 R 为互斥关系的时候，同时发生的概率最小，此时同时发生的概率为 0。

答案 A

5. The probability that both events A and B will occur is $\frac{1}{2}$. Which of the following values could be the probability that event A will occur?

Indicate <u>all</u> such values.

A. 0 B. $\frac{1}{4}$ C. $\frac{1}{2}$ D. $\frac{3}{4}$ E. 1

题干翻译

事件 A 和 B 都发生的概率是 $\frac{1}{2}$。下列哪个值可能是事件 A 发生的概率？

解题思路

概率的范围为 $0 \leqslant p \leqslant 1$。

因为 $P(A \cap B) = \frac{1}{2}$，

因此 $P(A) \geqslant P(A \cap B)\left(=\frac{1}{2}\right)$。

故 $\frac{1}{2} \leqslant P(A) \leqslant 1$。

答案 CDE

6. The probability that event R occurs is 0.25, and the probability that event T occurs is 0.20. If these events cannot both occur, what is the probability that neither of them occurs?

A. 0.95　　　　B. 0.55　　　　C. 0.50　　　　D. 0.45　　　　E. 0.05

题干翻译

事件 R 发生的概率是 0.25，事件 T 发生的概率是 0.20。如果这些事件不能同时发生，那么它们都不发生的概率是多少？

解题思路

根据题干关键词 cannot both occur 可知，事件 R 和 T 为互斥事件。

$$P(R \text{ 和 } T \text{ 都不发生}) = 1 - P(R \cup T),$$
$$= 1 - (0.2 + 0.25),$$
$$= 0.55。$$

答案　B

4.5 数据分布、随机变量和概率分布

基本词汇

random variable 随机变量　　　　　　standard normal distribution 标准正态分布
normal distribution 正态分布

数据分布

相对频率的表格或者柱状图是常被用来展现数据分布的方式。在柱状图中，条形的面积说明了数据集中的区域。

下图是对 800 个电子设备寿命的统计。因为寿命有许多不同的值，所以我们划定了 50 个区间，区间长度为 10 小时：601～610 小时，611～620 小时，…，1091～1100 小时。下图显示的是每个区间的相对频率。

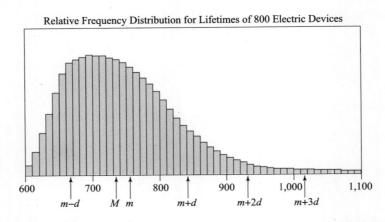

柱形图的顶部看起来相对平滑，像一条曲线。通常，呈现大量数据并分成许多连续的柱状图，其形状都十分平滑。因此，用接近条形顶部的曲线对数据分布进行建模，这个模型能体现数据分布的趋势。

柱状图中代表相对频率的条形面积之和为 1。尽管柱状图横轴上的单位会有变化，我们可以调整纵轴的单位使得条形面积总和为 1。调整之和，代表数据分布的曲线下方的面积也为 1，这条曲线叫作分布曲线，或者密度曲线或者频率曲线。

中位数把一组数据分成了上半部分和下半部分。在柱状图中，中位数左侧的条形面积之和等于中位数右侧的条形面积之和。

随机变量

随机变量是指随机事件的数量表现。随机事件数量化的好处是可以用数学分析的方法来研究随机现象。

正态分布

正态曲线呈钟形，两头低，中间高，左右对称，因其曲线呈钟形，因此人们又经常称之为钟形曲线。下图为平均数为 m，标准差为 d 的正态分布图。

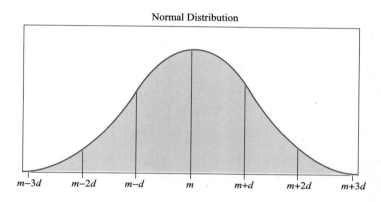

关于正态分布有以下几个性质：

- 平均数、中位数、众数为同一个数。
- 平均数把数据分为对称的两组。
- 大约 $\frac{2}{3}$ 的数据与平均数的偏离不超过 1 个标准差。
- 几乎所有的数据与平均数的偏离不超过 2 个标准差。

标准正态分布是一种特殊的正态分布，它的均值为 0，标准差为 1，如下图。

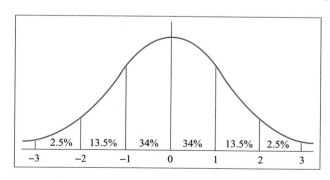

例题

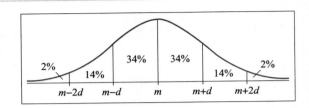

The figure above shows a normal distribution with mean m and standard deviation d, including approximate percents of the distribution corresponding to the six regions shown.

A survey of 5,500 book readers found that the number of books read per year was approximately normal distributed with mean 19.0 and standard deviation 2.0. Which of the following is the best description of the numbers of books read per year by the 880 book readers who read the most books?

A. 17 or more books
B. 19 to 21 books
C. 21 or more books
D. 21 to 23 books
E. 23 or more books

题干翻译

上图显示了具有平均值 m 和标准偏差 d 的正态分布，包括对应的所示 6 个区域分布的近似百分比。

一项针对 5500 名读者的调查发现，每年阅读的书籍数量大致呈正态分布，平均值为 19.0，标准差为 2.0。以下哪一项是对阅读书籍最多的 880 名读者每年阅读书籍数量的最佳描述？

解题思路

在正态分布图中，读书数量最多的 880 名读者占据了全部数据的 $\frac{880}{5500} = 16\%$，

因此，读书数量最多的 880 名受访者分布在 $m+d$ 之外（14% + 2%），

故他们的平均读书本数在 $m+d = 21$ 及以上。

答案 C

4.6 图表题

图表题考查的是考生读图、读文字的能力。这类题一般是给一张或两张表，然后让考生基于表格内容回答 3 个相关的问题，对应的知识点集中在本章节。在当前 GRE 考试中，图表题出现在第一个数学板块中，对应的题号为第 6 到第 8 题。

Questions 1–3 are based on the following data.

Internet Users in Seven Latin American Countries 1999 and 2003

Country	Number of Users (millions)		Percent of Population	
	1999	2003	1999	2003
Argentina	0.7	4.0	1.8%	10.3%
Brazil	5.8	20.1	3.3%	11.1%
Chile	0.3	1.4	1.9%	9.1%
Colombia	0.6	2.5	1.4%	5.8%
Mexico	1.0	4.8	1.0%	4.4%
Peru	0.2	1.0	0.7%	3.4%
Venezuela	0.3	1.4	1.4%	5.5%

* Numbers for the year 2003 are predicted values.

1. In 1999 the number of Internet users in Brazil was how many million greater than the total number of Internet users in the other six countries combined?

A. 2.1 B. 2.4 C. 2.7 D. 3.0 E. 3.3

题干翻译

1999 年，巴西的互联网用户数量比其他 6 个国家的互联网用户数量的总和多多少百万？

解题思路

根据图表可知，巴西 1999 年的互联网用户数量 = 5.8 million，

其他 6 个国家的互联网用户数量总和 = 0.7 + 0.3 + 0.6 + 1 + 0.2 + 0.3 = 3.1 million，

因此，巴西互联网用户数量 – 其他国家互联网用户数量总和 = 5.8 – 3.1 = 2.7 million。

答案　C

2. For the seven countries, what is the range of the numbers of Internet users predicted for the year 2003?

A. 16.4 million   B. 17.1 million   C. 17.6 million

D. 18.7 million   E. 19.1 million

题干翻译

对这 7 个国家来说，2003 年所预测的互联网用户数量的极差是多少？

解题思路

根据表格可知，这 7 个国家中，2003 年互联网用户数量最多的为巴西 20.1 million，2003 年互联网用户数量最少的为秘鲁 1 million。

极差 = 最大值 – 最小值 = 20.1 – 1 = 19.1 million。

答案   E

3. What is the predicted percent increase in the number of Internet users in Peru from 1999 to 2003?

A. 300%   B. 350%   C. 400%   D. 450%   E. 500%

题干翻译

从 1999 年到 2003 年，秘鲁互联网用户数量的预测增长百分比是多少？

解题思路

根据表格可知，秘鲁 1999 年互联网用户数量为 0.2 million，2003 年互联网用户数量为 1 million。

秘鲁 1999~2003 年互联网用户数量增长百分比 $= \dfrac{1-0.2}{0.2} = 400\%$。

答案   C

Questions 4–6 are based on the following data.

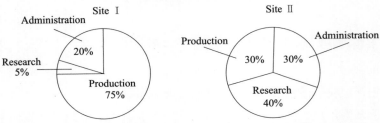

DISTRIBUTION OF EMPLOYEES OF COMPANY Y

Note: Each employee of Company Y works at one of the two sites in one of the three departments shown.

4. If 250 employees in administration were transferred from Site II to Site I and no other changes in employment occurred, approximately what percent of the employees at Site I would work in administration?

    A. 26%   B. 33%   C. 40%   D. 49%   E. 53%

题干翻译

如果250名行政部门的员工从工厂II转移到工厂I，并且没有发生其他雇用变化，那么工厂I的员工中大约有百分之多少将会在行政部门工作？

解题思路

根据图表可知，Site II 有1530名员工，Site I 有760名员工。

其中 Site I 之前的行政人员 = 760 × 20% = 152 人。

在 Site II 250名行政部门员工转到 Site I 之后，Site I 共 760 + 250 = 1010 名员工，Site I 的行政人员 = 250 + 152 = 402 人。

因此，转移后 Site I 的行政人员的占比 = $\frac{402}{1010} \approx 40\%$。

答案 C

5. The ratio of the number of production workers at Site I to the number of production workers at Site II is most nearly equal to

    A. 2 to 5   B. 4 to 5   C. 5 to 2   D. 5 to 3   E. 5 to 4

题干翻译

工厂I生产部门的工人人数与工厂II生产部门的工人人数的比率几乎等于多少？

解题思路

根据图表可知，Site I 生产部门的工人人数 = 760 × 75% = 570 人，

Site II 的生产工人人数 = 1530 × 30% = 459 人，

两者之比 = 570 : 459 ≈ 1.24，最接近 5 : 4。

答案 E

6. At Site II, if the average (arithmetic mean) salaries of employees in production, administration, and research are x, y, and z dollars, respectively, what is the average salary, in dollars, of all employees at Site II?

    A. $\dfrac{x+y+z}{3}$

    B. $0.3x + 0.3y + 0.4z$

    C. $\dfrac{0.3x + 0.3y + 0.4z}{3}$

D. $\dfrac{0.3x + 0.3y + 0.4z}{1,530}$

E. $(0.3)(1,530)x + (0.3)(1,530)y + (0.4)(1,530)z$

题干翻译

对于工厂 II，如果生产、管理和研究部门员工的平均（算术平均值）工资分别为 x、y 和 z 美元，那么工厂 II 所有员工的平均工资是多少美元？

解题思路

根据图表和题干条件可知，

总工资为 $1530 \times 30\% x + 1530 \times 30\% y + 1530 \times 40\% z$，

总人数为 1530，

因此，平均工资 $= \dfrac{总工资}{总人数} = \dfrac{1530 \times 30\% x + 1530 \times 30\% y + 1530 \times 40\% z}{1530} = 0.3x + 0.3y + 0.4z$。

答案 B

Questions 7–9 are based on the following data.

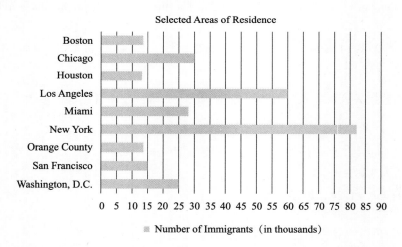

The graph shows the numbers of all immigrants admitted to the United States in year Y who intended to reside in nine selected areas.

7. For year Y, approximately what was the median of the numbers of immigrants admitted to the United States who intended to reside in the nine areas shown?

A. 15,000　　B. 20,000　　C. 25,000　　D. 30,000　　E. 35,000

题干翻译

在 Y 年，打算居住在所示 9 个地区的美国移民人数的中位数大约是多少？

解题思路

对于 9 个地区而言，中位数应该是从小到大排序后，移民人数第 5 的地区。

读图可知，移民人数第 5 多的地区为 Washington, D. C.，其数量为 25000。

答案　C

8. One immigrant is to be randomly chosen from the immigrants admitted to the United States in year Y who intended to reside in one of three areas: Chicago, Los Angeles, and New York. Approximately what is the probability that the immigrant chosen will be one who intended to reside in Chicago?

   A. $\dfrac{1}{172,600}$　　　B. $\dfrac{1}{31,000}$　　　C. $\dfrac{1}{17}$　　　D. $\dfrac{3}{17}$　　　E. $\dfrac{1}{3}$

题干翻译

从 Y 年获准进入美国的移民中随机选择一名移民，他打算居住在芝加哥、洛杉矶和纽约这 3 个地区之一。被选中的移民打算居住在芝加哥的概率大约是多少？

解题思路

通过图表可知，Chicago 的移民人数大约为 30 thousand，Los Angeles 的移民人数大约为 60 thousand，New York 的移民人数大约为 82 thousand。

因此，被选中的移民居住在 Chicago 的概率 $= \dfrac{30}{30+60+82} \approx \dfrac{3}{17}$。

答案　D

9. Based on the information given, which of the following statements about the numbers of immigrants who intended to reside in the 9 areas shown are true?

   Indicate <u>all</u> such statements.

   A. The average (arithmetic mean) of the numbers of immigrants for the 9 areas is greater than the number of immigrants who intended to reside in Washington, D. C.
   B. The range of the numbers of immigrants for the 9 areas is greater than the number of immigrants who intended to reside in Los Angeles.
   C. For the 9 areas, the standard deviation of the least 5 numbers of immigrants is less than the standard deviation of the greatest 4 numbers of immigrants.

题干翻译

根据上述的信息，下列哪项关于打算居住在所示 9 个地区的移民人数的陈述是正确的？

解题思路

根据图表可知，9 个地区的移民平均数 $= \dfrac{13+14+14+15+25+28+30+60+82}{9} \approx 31.2$ thousand，大于 Washington, D.C. 的移民人数 25 thousand，A 选项正确。

9 个地区移民人数的极差 $= 82 - 13 = 69$ thousand，大于 Los Angeles 的移民人数 60 thousand，B 选项正确。

最小的 5 个数字为 13，14，14，15，25；最大的 4 个数字为 28，30，60，82。根据观察可知，最大的 4 个数字的离散程度更大，因此标准差更大，C 选项正确。

答案 A B C

Questions 10–12 are based on the following data.

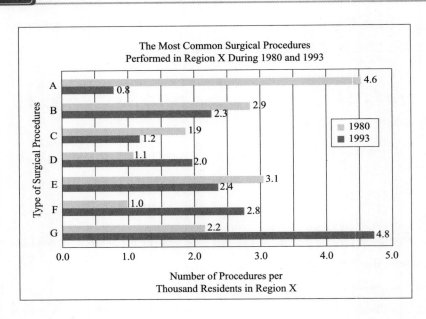

10. In 1980 what was the median number of surgical procedures performed per thousand residents for the seven types of surgical procedures shown?

A. 1.1　　B. 1.9　　C. 2.2　　D. 2.4　　E. 2.6

题干翻译

1980 年，在所示的 7 种外科手术中，每千名居民进行的外科手术的中位数是多少？

解题思路

对于 7 种外科手术,中位数为从小到大排序后第 4 位对应的数值。1980 年,从小到大排序后,第 4 位为 G,对应的数值为 2.2。

答案 C

11. For which type of surgical procedures did the number of procedure per thousand residents in region X increase by the greatest percent from 1980 through 1993?

A. A   B. B   C. D   D. F   E. G

题干翻译

从 1980 年到 1993 年,X 区每千名居民中哪种类型的外科手术的数量增长最快?

解题思路

D 区域增长百分比 $= \dfrac{2-1.1}{1.1} \approx 82\%$,

F 区域增长百分比 $= \dfrac{2.8-1}{1} = 180\%$,

G 区域增长百分比 $= \dfrac{4.8-2.2}{2.2} \approx 118\%$,

因此,F 区域增长百分比最大。

答案 D

12. For how many of the seven types of surgical procedures did the number of procedures per thousand residents in region X increase or decrease by more than 50 percent from 1980 through 1993?

A. Two   B. Three   C. Four   D. Five   E. Six

题干翻译

从 1980 年到 1993 年,X 区每千名居民进行的 7 种外科手术中有多少区域增加或减少了 50% 以上?

解题思路

A 区域减少百分比 $= \dfrac{4.6-0.8}{4.6} \approx 83\% > 50\%$,

B 区域减少百分比 $= \dfrac{2.9-2.3}{2.9} \approx 21\% < 50\%$,

C 区域减少百分比 $= \dfrac{1.9-1.2}{1.9} \approx 37\% < 50\%$,

D 区域增长百分比 $= \dfrac{2-1.1}{1.1} \approx 82\% > 50\%$,

E 区域减少百分比 = $\dfrac{3.1-2.4}{3.1} \approx 23\% < 50\%$,

F 区域增长百分比 = $\dfrac{2.8-1}{1} = 180\% > 50\%$,

G 区域增长百分比 = $\dfrac{4.8-2.2}{2.2} \approx 118\% > 50\%$。

因此，X 区每千名居民进行的 7 种外科手术中增加或减少了 50% 以上的区域有 A，D，F，G 4 个区域。

答案　C

Questions 13－15 are based on the following data.

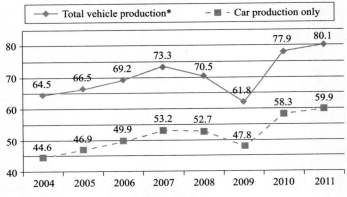

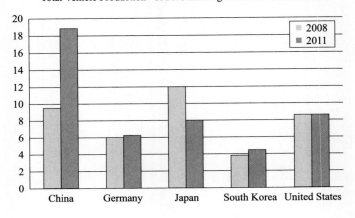

13. For the years 2008 and 2011 combined, the total number of vehicles produced in Japan was approximately what percent of the total number of vehicles produced globally?

A. 10%　　B. 13%　　C. 16%　　D. 19%　　E. 22%

题干翻译

2008 年和 2011 年，日本生产的交通工具总数约占全球生产的交通工具总数的百分之几？

解题思路

根据图表可知：

2008 年和 2011 年全球生产的 vehicle 数量 = 70.5 + 80.1 = 150.6 million（上图），

2008 年和 2011 年日本生产的 vehicle 数量 = 12 + 8 = 20 million（下图），

日本生产的 vehicle 数量在全球的占比 = $\frac{20}{150.6} \approx 13\%$。

答案　B

14. In 2011, if the ratio of the number of cars produced in China to the total number of vehicles produced in China was equal to the corresponding ratio globally, approximately how many vehicles other than cars were produced in China in 2011?

A. 5 million　　B. 9 million　　C. 11 million　　D. 14 million　　E. 19 million

题干翻译

2011 年，如果中国生产的汽车数量与中国生产的交通工具总数的比率等于全球相应的比率，那么 2011 年中国生产了多少辆汽车以外的交通工具？

解题思路

设 2011 年中国生产了 x 辆汽车。

根据题干条件和图表信息可知：

$\frac{\text{China's cars}}{\text{China's vehicles}} = \frac{x}{18.8} = \frac{\text{global cars}}{\text{global vehicles}} = \frac{59.9}{80.1}$，

$x \approx 14$，

因此，2011 年中国生产的其他交通工具数量 = 18.8 − 14 = 4.8 ≈ 5 million。

答案　A

15. A vehicle is to be selected random from all the vehicles produced globally in 2008. Approximately what is the probability that the vehicle will be one that was produced in China or Japan?

A. 0.03　　B. 0.15　　C. 0.30　　D. 0.35　　E. 0.40

题干翻译

从 2008 年全球生产的所有交通工具中随机选择一辆交通工具。该车辆在中国或日本生产的概率大约有多大?

解题思路

概率 = $\dfrac{中国 + 日本}{总量} = \dfrac{9.5 + 12}{70.5} \approx 0.3$。

答案 C

Questions 16–18 are based on the following data.

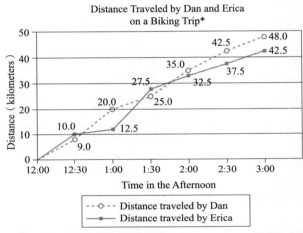

*Dan and Erica started traveling from the same place at 12:00 in the afternoon and traveled in the same direction along the same route until 3:00 in the afternoon.

16. For which of the following time periods was Erica's average speed greater than Dan's average speed? Indicate all such time periods.

A. From 12:00 to 12:30   B. From 12:30 to 1:00   C. From 1:00 to 1:30
D. From 1:30 to 2:00   E. From 2:00 to 2:30   F. From 2:30 to 3:00

题干翻译

在以下哪个时间段,Erica 的平均速度高于 Dan 的平均速度?

解题思路

速度 = $\dfrac{距离}{时间}$,本题的速度体现在线段的斜率上,斜率越大(线段越陡峭),速度越大。

因此，要找 Erica 的平均速度高于 Dan 的平均速度的时间段，只要找 Erica 的斜率比 Dan 斜率大的线段即可。

符合要求的有：12:00 to 12:30 和 1:00 to 1:30。

答案　AC

17. For which of the following time periods was Dan's average speed closest to his average speed for the time period from 12:00 to 3:00?

A. From 12:00 to 12:30
B. From 12:30 to 1:00
C. From 1:00 to 1:30
D. From 1:30 to 2:00
E. From 2:00 to 2:30

题干翻译

在以下哪个时间段，Dan 的平均速度最接近他在 12:00 到 3:00 期间的平均速度？

解题思路

Dan 在 12:00～3:00 的平均速度 $=\dfrac{48}{3}=16$ km/h。

Dan 在 12:00～12:30 的平均速度 $=\dfrac{9}{0.5}=18$ km/h，与 16 km/h 相差 2 km/h。

Dan 在 12:30～1:00 的平均速度 $=\dfrac{11}{0.5}=22$ km/h，与 16 km/h 相差 6 km/h。

Dan 在 1:00～1:30 的平均速度 $=\dfrac{5}{0.5}=10$ km/h，与 16 km/h 相差 6 km/h。

Dan 在 1:30～2:00 的平均速度 $=\dfrac{10}{0.5}=20$ km/h，与 16km/h 相差 4 km/h。

Dan 在 2:00～2:30 的平均速度 $=\dfrac{7.5}{0.5}=15$ km/h，与 16km/h 相差 1km/h。

因此最接近 12:00 到 3:00 期间的平均速度的时间段为 2:00 to 2:30。

答案　E

18. Erica's average speed for the time period from 12:30 to 1:00 was how many kilometers per hour less than her average speed for the time period from 1:00 to 1:30?

A. 5.0
B. 12.5
C. 15.0
D. 25.0
E. 30.0

题干翻译

Erica 在 12:30 到 1:00 这段时间内的平均速度比她在 1:00 到 1:30 这段时间内的平均速度每小时低多少公里？

解题思路

Erica 在 12:30~1:00 的平均速度 $=\dfrac{2.5}{0.5}=5$ km/h。

Erica 在 1:00~1:30 的平均速度 $=\dfrac{15}{0.5}=30$ km/h。

$30-5=25$ km/h。

答案 D

Questions 19–21 are based on the following data.

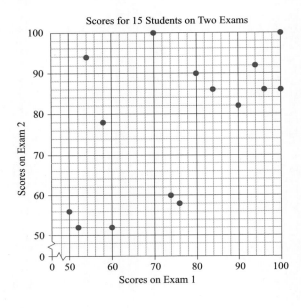

Note:

The two coordinates of each dot on the graph represent the two exam scores of a student, where each score is a whole number.

19. The median of the scores on exam 1 was how much higher or lower than the median of the scores on exam 2?

A. 10 lower   B. 6 lower   C. 2 higher   D. 4 higher   E. 8 higher

题干翻译

考试 1 分数的中位数比考试 2 分数的中位数高或低多少?

解题思路

共 15 个学生，因此中位数为从小到大排序后第 8 个学生的分数。

根据图表可知，exam 1 从小到大排序后第 8 个学生的成绩为 76，exam 2 从小到大排序后第 8 个学生成绩为中位数 86。因此，exam 1 的中位数比 exam 2 的中位数低了 10 分，即 10 lower。

答案 A

20. One of the 15 students had a total score of 182 on the two exams. What is the weighted mean of that student's two scores if the score on exam 1 has a weight of 40 percent and the score on exam 2 has a weight of 60 percent?

A. 89　　B. 90　　C. 91　　D. 92　　E. 93

题干翻译

15 名学生中有一名学生两次考试的总分为 182 分。如果考试 1 的分数权重为 40%，考试 2 的分数权重为 60%，那么该学生两次考试的分数的加权平均值是多少？

解题思路

散点图取数的时候，可以把横轴和纵轴看作直角坐标，利用直线来读数。

如下图所示，在 exam 1 和 exam 2 的成绩里面各对应一个点（82，100）和（100，82），连接这两个点可以得到一条直线，在这条直线上的点，成绩都是 182。我们发现图中只有一个点在这条直线上，就可以快速地确定这个点。

这个点的横坐标和纵坐标分别为 96 和 86，和为 182。再根据题目条件求得该学生两次考试的加权平均数为：$96 \times 0.4 + 86 \times 0.6 = 90$。

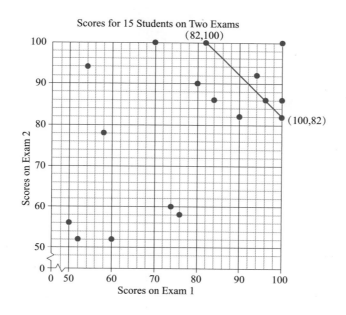

答案 B

21. A pair of students is to be chosen to write a certain report, and each of the chosen students must have a total score of 170 or higher on the two exams. How many different pairs of students meet this requirement?

A. 6　　　　B. 8　　　　C. 16　　　　D. 21　　　　E. 28

题干翻译

将选择一对学生写某份报告，每个被选中的学生必须在两次考试中获得170分或更高的总分。有多少对不同组合的学生符合这个要求？

解题思路

与上一题同理，我们可以把exam 1和exam 2里面170所在的点坐标（70，100）和（100，70）连接起来组成一条直线，如下图所示，在直线上以及直线上方的点，其横坐标与纵坐标的和肯定都是大于等于170的。我们可以从下图读出，一共有8个点符合要求，即有8个学生两次考试的总分大于等于170，在这8个学生中任取2个点，即$C_8^2 = 28$。

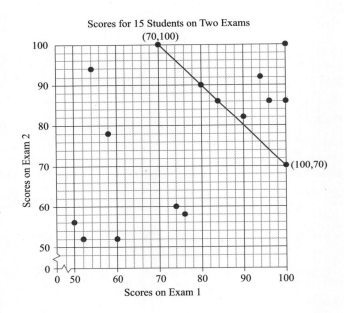

答案　E

4.7 应用题

应用题对考生的英文阅读能力有比较高的要求，需要考生将冗长的文字变成对应的数学表达式。GRE 数学常考的应用题有以下几种。

4.7.1 平均数问题

平均数问题的核心公式：平均数 = $\frac{总和}{总个数}$。

The table shows the prices of three brands of bread and the corresponding number of loaves sold yesterday at a local market. If the average (arithmetic mean) price per loaf of all the loaves sold yesterday was greater than $2.50, which of the following could be the value of n?
Indicate all such statements.

Brand	Price Per Loaf	Number of Loaves Sold
A	$4.00	12
B	$2.00	n
C	$1.00	8

A. 7   B. 9   C. 11   D. 13   E. 15

题干翻译

表格显示了昨天在当地市场出售的 3 种品牌面包的价格和相应的面包数量。如果昨天售出的所有面包的平均价格大于 2.50 美元，下列哪一项可能是 n 的值？

解题思路

平均数 = $\frac{总和}{总个数}$ = $\frac{4 \times 12 + 2n + 1 \times 8}{12 + n + 8}$ = $\frac{56 + 2n}{20 + n} > 2.5$，

$56 + 2n > 50 + 2.5n$，

$n < 12$，

因此，符合要求的选项有 ABC。

答案 ABC

4.7.2 物质混合问题

物质混合问题是把不同性质的物质混合在一起，通常考查浓度的百分比或者某一物质的占比等问题。解决这类问题，务必记住物质守恒，即物质混合前和混合后数量保持不变。在碰到物质混合问题时，好好读题，判断题目研究的是哪一种物质，然后将该物质混合前的数量表示出来，再将该物质混合后的数量表示出来，最后在中间联立等号，求解即可。

When 9 gallons of a solution that is 30 percent alcohol are mixed with 18 gallons of a solution that is x percent of alcohol, the resulting mixture is 25 percent alcohol. Which of the following is closest to the value of x?

A. 15　　　　B. 20　　　　C. 23　　　　D. 25　　　　E. 27

题干翻译

当 9 加仑含 30% 酒精的溶液与 18 加仑含 x% 酒精的溶液混合时，所得混合物中酒精浓度为 25%。下列哪一项最接近 x 的值？

解题思路

混合问题，物质混合前后数量不变。

混合前的酒精总量 $= 9 \times 30\% + 18x\%$，

混合后的酒精总量 $= (18 + 9) \times 25\%$，

$9 \times 30\% + 18x\% = (18 + 9) \times 25\%$，

$x = 22.5$，

最接近 x 的值的是 C 选项。

答案　C

4.7.3 路程问题

路程问题有相遇问题、追及问题等，路程问题的核心公式：距离 = 速度 × 时间。

1. 流水问题

船在江河里航行时，除了本身的前进速度外，还受到流水的推送或顶逆，在这种情况下计算船只的航行速度、时间和所行的路程，叫作流水问题。流水问题中 3 个量（速度、时间、路程）的关系在这里将要反复用到。

此外，流水行船问题还有以下两个基本公式：

顺流速度 = 船速 + 水速；

逆流速度 = 船速 − 水速。

例题

A boat traveled upstream a distance of 90 miles at an average speed of $(v-3)$ miles per hour and then traveled the same distance downstream at an average speed of $(v+3)$ miles per hour. If the trip upstream took half an hour longer than the trip downstream, how many hours did it take the boat to travel downstream?

A. 2.5　　　B. 2.4　　　C. 2.3　　　D. 2.2　　　E. 2.1

题干翻译

一艘船以每小时 $(v-3)$ 英里的平均速度向上游行驶 90 英里，然后以每小时 $(v+3)$ 英里的平均速度向下游行驶相同的距离。如果逆流而上比顺流而下多花了半个小时，那么船顺流而下需要多少个小时？

解题思路

逆流时间 − 顺流时间 = 0.5，

$\dfrac{距离}{逆流速度} - \dfrac{距离}{顺流速度} = 0.5$，

$\dfrac{90}{v-3} - \dfrac{90}{v+3} = 0.5$，

$90(v+3) - 90(v-3) = 0.5(v^2 - 9)$，

$v^2 = 1089$，

$v = 33$，

顺流时间 $= \dfrac{90}{33+3} = 2.5$。

答案　A

2. 追及问题

追及问题指两个运动物体在不同地点同时出发（或者在同一地点而不是同时出发，或者在不同地点又不是同时出发）作同向运动，在后面的物体行进速度要快些，在前面的物体行进速度要慢些，在一定时间之内，后面的物体追上前面的物体。

> **例题**
>
> Jessica and Anna go running together along the same line. Jessica started 30 seconds later than Anna because she had to tie her shoelaces. Jessica runs 5 feet/second faster than Anna, and Jessica catches up with Anna at a distance of 1800 feet from the starting point. How many seconds did Jessica spend chasing?
>
> **题干翻译**
> Jessica 和 Anna 两个人沿着同一条线路一起去跑步。因为 Jessica 要系鞋带，所以比 Anna 晚了 30 秒出发。已知 Jessica 比 Anna 的速度快 5 feet/second，并且在距离出发点 1800 feet 的位置 Jessica 追上了 Anna。请问 Jessica 花了多少秒追上？
>
> **解题思路**
> 设 Jessica 追的时间为 t 秒，则 Anna 跑步时间为 $t+30$ 秒。
> 根据题干信息可知：
> Jessica 的速度 − Anna 的速度 = 5，
> $$\frac{距离}{\text{Jessica 跑步时间}} - \frac{距离}{\text{Anna 跑步时间}} = 5,$$
> $\frac{1800}{t} - \frac{1800}{t+30} = 5,$
> $1800(t+30) - 1800t = 5t(t+30),$
> $1800 \times 30 = 5t(t+30),$
> $t(t+30) = 10800,$
> $t^2 + 30t - 10800 = 0,$
> $(t-90)(t+120) = 0,$
> $t = 90$（时间必须是非负数）。
>
> **答案** 90 秒

3. 相遇问题

相遇问题是指两个物体从两地同时出发，面对面相向而行，经过一段时间，两个物体必然会在途中相遇。相遇问题的基本公式：速度 × 时间 = 路程。

从出发到相遇的时间是相遇时间，从出发到相遇一起走的路程是相遇路程，单位时间内一起走的路程是两个物体的路程和。注意：必须是同时同步的。

Jeff and Dennis ride bicycles in opposite directions from the two places at the same time. Jeff rides 15 kilometers per hour and Dennis rides 13 kilometers per hour. If the distance between the two places is 112 kilometers, how many hours will it take for Jeff and Dennis to meet?

题干翻译

Jeff 和 Dennis 二人同时从两地骑自行车相向而行，Jeff 的骑行速度为 15 千米/小时，Dennis 的骑行速度为 13 千米/小时。如果两地距离为 112 千米，那么 Jeff 和 Dennis 多少小时后会相遇？

解题思路

相遇时间 = $\dfrac{\text{距离}}{\text{Jeff 的骑行速度} + \text{Dennis 的骑行速度}} = \dfrac{112}{15+13} = 4$。

答案 4 小时

4. 平均速度问题

注意，平均速度并不一定等于速度的平均值。正确的思路：平均速度 = $\dfrac{\text{总距离}}{\text{总时间}}$，因此涉及平均速度的题目，我们应该找出总距离和总时间，再将它们代入公式就行！

Ellen drove the 240 miles from Springfield to Greenville at a rate of 40 miles per hour. Then he returned along the same route at a rate of 60 miles per hour. What was his average speed for the entire trip?

A. 42 miles per hour B. 48 miles per hour C. 50 miles per hour
D. 54 miles per hour E. 56 miles per hour

题干翻译

Ellen 以每小时 40 英里的速度从 Springfield 驱车 240 英里到达 Greenville。然后他以每小时 60 英里的速度沿着同一条路线返回。他整个旅程的平均速度是多少？

解题思路

先计算总距离：本题中的总距离为往返路程之和 S = 240 × 2 = 480 miles。

再计算总时间：本题中的总时间是两段路程的总时间，因为两段路程速度不同，所以时间需要分段计算。

$t = t_1 + t_2 = \dfrac{240}{40} + \dfrac{240}{60} = 10$ hours，

平均速度 = $\dfrac{\text{总距离}}{\text{总时间}} = \dfrac{480}{10} = 48$ miles per hour。

答案 B

4.7.4 生产问题

在解生产问题的时候，大家要牢记公式：
工作任务量 = 工作效率 × 工作时间。

在题目中没有明确说明工作量的时候，我们一般默认工作量为 1，然后需要注意，只有工作效率可以叠加，工作时间是不能叠加的。比如，A 做一件事需要 4 小时，B 做同样一件事需要 5 小时，你不能说两个人一起做这件事需要 9 小时。两个人一起做同一件事，明显速度会变快，时间会变短才对！所以做题的时候，告诉我们单独的工作时间，然后问一起的工作时间或者告诉我们一起的工作时间，让我们换算单独的工作时间，我们在做加减的时候，只能是工作效率相加减。

例题 01.

Two water faucets are used to fill a certain tank. Running individually at their respective constant rates, these faucets fill the empty tank in 12 minutes and 20 minutes, respectively. If no water leaves the tank, how many minutes will it take for both faucets running simultaneously at their respective rates to fill the empty tank?

A. 16.0　　　　B. 10.5　　　　C. 8.0　　　　D. 7.5　　　　E. 6.0

题干翻译
两个水龙头用来给一个水箱注水。这些水龙头以各自恒定的速度单独运行，分别在 12 分钟和 20 分钟内注满空水箱。如果水箱不漏水，两个水龙头以各自的速度同时运行需要多少分钟才能注满空水箱？

解题思路
根据题干中告诉我们的时间换算出各自的工作效率，

$$工作效率 = \frac{工作量}{工作时间}。$$

题目中没有明确说明工作量的时候我们一般将工作量按 1 来计算：两个水龙头各自的工作效率为 $\frac{1}{12}$ 和 $\frac{1}{20}$。

将各自的工作效率相加，算出一起的工作效率：$\frac{1}{12} + \frac{1}{20} = \frac{8}{60} = \frac{2}{15}$。

再用工作量除以工作效率，算出一起完成任务所需的工作时间：

$$\frac{1}{\frac{2}{15}} = \frac{15}{2} = 7.5。$$

答案　D

Faucets X and Y dispense water at their respective constant rates. Dispensing simultaneously, faucets X and Y fill an empty tub in 3 minutes. Faucet X, dispensing alone, takes $\frac{3}{4}$ as long as faucet Y, dispensing alone, to fill the empty tub. Let x and y be the respective times in minutes, that it takes faucets X and Y, each dispensing alone, to fill the empty tub. What are the values of x and y?

A. $x = 3$ and $y = 4$
B. $x = 4.5$ and $y = 6$
C. $x = 5.25$ and $y = 7$
D. $x = 6$ and $y = 8$
E. $x = 7$ and $y = 9.3$

题干翻译

水龙头 X 和 Y 以各自恒定的速度供水。水龙头 X 和 Y 同时供水，在 3 分钟内注满一个空浴缸。水龙头 X 单独注满空浴缸所花时间是水龙头 Y 单独注满空浴缸所花时间的 $\frac{3}{4}$。x 和 y 分别表示以分钟为单位，水龙头 X 和水龙头 Y 将空浴缸注满水分别所需的时间。x 和 y 的值是多少？

解题思路

根据题干中告诉我们的时间换算出各自的工作效率，题目中没有明确说明工作量的时候我们一般将工作量按 1 来计算。X 与 Y 两个水龙头的各自的工作效率为 $\frac{1}{x}$ 和 $\frac{1}{y}$。

将各自的工作效率相加，算出一起的工作效率：$\frac{1}{x} + \frac{1}{y} = \frac{1}{3}$。

根据题干中的条件 "Faucet X, dispensing alone, takes $\frac{3}{4}$ as long as faucet Y" 可知 $x = \frac{3}{4}y$。

联立方程组：
$$\begin{cases} \frac{1}{x} + \frac{1}{y} = \frac{1}{3}, \\ x = \frac{3}{4}y, \end{cases}$$

可解出 $x = 5.25$，$y = 7$。

答案 C

4.7.5 投资问题

----- 基本词汇 -----

simple interest 单利   revenue 收入   profit 利润
compound interest 复利   expense/cost 成本   gross profit 毛利

1. 利息问题

利息问题我们需要掌握两个考点。

(1) 单利和复利

单利：本息和 = 本金 + 本金 × 利率 × 存期 = 本金（1 + 利率 × 存期）。

复利：本息和 = 本金 ×（1 + 利率）^{存期}。

注意，单利和复利的算法是不一样的，因此审题时务必注意题干问的是单利问题还是复利问题，然后代入相应的公式。

(2) 存期期限

计算利息的时候，特别是复利问题的时候，务必看清楚题干的存款方法是"一年一期""半年一期"还是"一季度一期"。

例1：小明将10000元存入年利率为6%的复利账户，存了1年，1年之后小明的账户里有多少钱？

$$10000 \times (1 + 6\%)。$$

例2：小明将10000元存入年利率为6%的复利账户，存了1年，半年为一期，1年之后小明的账户里有多少钱？

$$10000 \times \left(1 + \frac{6\%}{2}\right)^2 = 10000 \times (1.03)^2。$$

注意，如果题干中给的是年利率，复利方式是按半年复利一次，我们应该基于年利率换算出其对应的半年利率，官方命题组默认半年利率 = $\frac{年利率}{2}$；同理，如果题干中给的是年利率，复利方式是按季度复利，官方命题组默认季度利率 = $\frac{年利率}{4}$。

Number of Dollars Invested	Simple Annual Interest Rate
12,000	3.0%
20,000	1.8%
x	4.5%
y	3.6%

The table shows four different amounts of money invested on the same day at different simple annual interest rates. If each investment earns the same amount of interest for the first year after the investment is made, what is the value of $y - x$?

A. 2,000　　　　B. 4,000　　　　C. 6,000　　　　D. 8,000　　　　E. 10,000

题干翻译

该表显示了在同一天以不同的年度单利投资的 4 种不同金额的资金。如果每项投资在投资后的第一年获得的利息相同，那么 $y-x$ 的值是多少？

解题思路

单利利息 = 本金 × 利率。

$12000 \times 3\% = x \cdot 4.5\% = y \cdot 3.6\%$，

解得 $x = 8000$，$y = 10000$，

$y - x = 2000$。

答案　A

例题 02.

Paul's family put m dollars in a new savings account on May 2, 1990, and put the same number of dollars in the account on May 2, 1991, and again on May 2, 1992. If the annual interest rate on this account was 4 percent compounded annually and there were no other deposits to the account or withdrawals from the account, which of the following represents the total number of dollars in the account on May 2, 1993, just after interest had been compounded for the third time, in terms of m?

A. $m(1.04)$

B. $3m(1.04)$

C. $m(1.04) + 2m(1.04) + 3m(1.04)$

D. $m(1.04) + m(1.04)^2 + m(1.04)^3$

E. $m(1.04) + m^2(1.04)^2 + m^3(1.04)^3$

题干翻译

Paul 的家庭于 1990 年 5 月 2 日将 m 美元存入一个新的储蓄账户，并于 1991 年 5 月 2 日和 1992 年 5 月 2 日将相同数量的美元存入该账户。如果这个账户的年利率是每年复利 4%，并且没有其他存款或提款，下列哪一项代表 1993 年 5 月 2 日账户中在第三次复利之后的美元总数？

解题思路

复利本息和 = 本金 × $(1 + 利率)^{存期}$。

1990 年存入账户金额 m 在 1993 年的本息和 = $m(1+4\%)^3 = m(1.04)^3$（复利三次），

1991 年存入账户金额 m 在 1993 年的本息和 = $m(1+4\%)^2 = m(1.04)^2$（复利两次），

1992 年存入账户金额 m 在 1993 年的本息和 = $m(1+4\%) = m(1.04)$（复利一次），

因此，1993 账户中总的本息和 = $m(1.04) + m(1.04)^2 + m(1.04)^3$。

答案　D

例题 03

A 2-year certificate of deposit is purchased for k dollars. If the certificate earns interest at an annual rate of 6 percent compounded quarterly, which of the following represents the value, in dollars, of the certificate at the end of the 2 years?

A. $(1.06)^2 k$ B. $(1.06)^8 k$ C. $(1.015)^2 k$

D. $(1.015)^8 k$ E. $(1.03)^4 k$

题干翻译

用 k 美元购买一张 2 年期的存单。如果这张存单的年利率为 6%，按每季度复利，下列哪个选项代表该存单在两年后的金额？

解题思路

年利率为 6%，则季度利率为 $\dfrac{6\%}{4} = 1.5\%$。

存了 2 年，1 年 4 个季度，1 个季度为 1 期，说明存了 8 期。

因此，两年后账户的本息和 $= k \times (1 + 1.5\%)^8$。

答案 D

2. 利润成本问题

$$利润 = 收入 - 成本。$$

例题

A dealer's gross profit on the sale of a car is defined as the selling price of the car minus its cost to the dealer. A car dealer sold n cars for s dollars each. If the dealer's total gross profit was p dollars, what was the dealer's average (arithmetic mean) cost per car, in dollars?

A. $\dfrac{s-p}{n}$ B. $s - \dfrac{p}{n}$ C. $\dfrac{s}{n} - p$ D. $p - \dfrac{s}{n}$ E. $\dfrac{sn}{p}$

题干翻译

经销商销售汽车的毛利被定义为汽车的销售价格减去经销商的成本。一个汽车经销商以每辆 s 美元的价格卖了 n 辆汽车。如果经销商的总毛利是 p 美元，那么经销商每辆车的平均成本是多少美元？

解题思路

利润 = 收入 – 成本。

总收入 = 售价 × 数量 = $s \times n$ = sn，

总成本 = 总收入 – 利润 = $sn - p$，

平均成本 = $\dfrac{总成本}{数量} = \dfrac{sn-p}{n} = s - \dfrac{p}{n}$。

答案 B

4.7.6 钱的计算问题

钱的计算问题经常考查"阶梯性定价",即根据不同的购买数量或者消费水平来制定不同的价格。我们生活中的打车费、个人所得税、运费等都属于阶梯性定价。阶梯性定价的最终费用等于各段费用的总和。

In 2013, Jean had taxable income equal to $80,500. Her income tax for 2013 was assessed as 10 percent of her taxable income up to $8,925, plus 15 percent of her taxable income in excess of $8,925 and up to $36,250, plus 25 percent of her taxable income in excess of $36,250. What was the total amount assessed for Jean's income tax for 2013, rounded to the nearest dollar?

A. $15,161   B. $16,054   C. $17,891   D. $20,125   E. $22,540

题干翻译

2013年,Jean 的应税收入为80500美元。她2013年的所得税,在其不超过8925美元部分为应税收入的10%,加上其超过8925美元且不超过36250美元部分的应税收入的15%,加上其超过36250美元的应税收入的25%。Jean 2013年的所得税总额是多少(四舍五入到美元)?

解题思路

根据题干可知,Jean 在2013年的所得税由3个部分构成,分别是不超过8925美元的所得税,8925到36250美元的所得税和超过36250美元的所得税。

Jean 2013 的所得税 $= 8925 \times 10\% + (36250 - 8925) \times 15\% + (80500 - 36250) \times 25\%$,

$\qquad = 892.5 + 4098.75 + 11062.5$,

$\qquad = 16053.75$,

$\qquad \approx 16054$。

答案 B

第三篇

模考套题训练

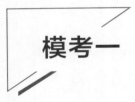

SECTION 1
Quantitative Reasoning

Time—21 minutes
12 Questions

Question 1

Quantity A
$(a-b)+(c-d)$

Quantity B
$(a+c)-(b+d)$

A. Quantity A is greater.
B. Quantity B is greater.
C. The two quantities are equal.
D. The relationship cannot be determined from the information given.

Question 2

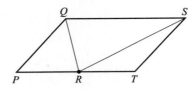

PQST is a parallelogram and R is the midpoint of side PT.

Quantity A
The area of triangular region PQR

Quantity B
The area of triangular region RST

A. Quantity A is greater.
B. Quantity B is greater.
C. The two quantities are equal.
D. The relationship cannot be determined from the information given.

Question 3

Quantity A	Quantity B
The least value of y for which $-5 \leq 1 - 2y \leq 9$	-2

A. Quantity A is greater.

B. Quantity B is greater.

C. The two quantities are equal.

D. The relationship cannot be determined from the information given.

Question 4

$|x| \leq 6$ and $|y| \leq 4$

x and y are integers, where $x \neq 0$. M is the greatest possible value of $\left|\dfrac{y}{x}\right|$.

Quantity A	Quantity B
M	1

A. Quantity A is greater.

B. Quantity B is greater.

C. The two quantities are equal.

D. The relationship cannot be determined from the information given.

Question 5

$2x + y = 5$

If integers x and y satisfy the equation shown, then $(9^x)(3^y) =$

A. 27   B. 81   C. 243   D. 729   E. 2,187

Questions 6–8 are based on the following data.

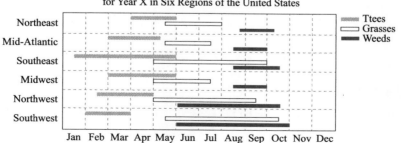

Pollen Production Periods of Treees, Grasses, and Weeds for Year X in Six Regions of the United States

6. Which of the following is closest to the median length of the time periods, in months, during which trees produced pollen in the six regions?

   A. 2.0 B. 2.4 C. 2.8 D. 3.0 E. 3.2

7. For approximately what fraction of year X was pollen produced by both grasses and weeds in the Southwest?

   A. $\dfrac{9}{24}$ B. $\dfrac{11}{24}$ C. $\dfrac{12}{24}$ D. $\dfrac{14}{24}$ E. $\dfrac{20}{24}$

8. For year X, which of the following is closest to the total length, in months, of the time period during which no pollen was produced by trees or grasses in any of the six regions?

   A. 0.5 B. 1.0 C. 2.0 D. 2.5 E. 3.0

Question 9

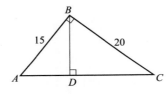

In the figure shown, what is the length of line segment BD?

_____

Question 10

$$-\dfrac{1}{2},\ \left(-\dfrac{1}{2}\right)^2,\ \left(-\dfrac{1}{2}\right)^3,\ \left(-\dfrac{1}{2}\right)^4$$

What is the range of the four numbers listed above?

A. $\dfrac{1}{8}$ B. $\dfrac{1}{4}$ C. $\dfrac{3}{8}$ D. $\dfrac{1}{2}$ E. $\dfrac{3}{4}$

Question 11

Which of the following is equal to $(6)(14^8) + (15)(14^7)$?

A. $(21)(14^7)$ B. $(35)(14^7)$ C. $(84)(14^7)$ D. $(90)(14^7)$ E. $(99)(14^7)$

Question 12

Mr. Thomas gave a chemistry test to 25 students and assigned each student a score. Of the 25 students, 12 students received scores that were greater than 80.

Which of the following statements individually provides sufficient additional information to determine the median of the 25 scores?

Indicate all such statements.

A. The average (arithmetic mean) of the 25 scores was 80.
B. One student received a score of 80.
C. Twelve students received scores that were less than or equal to 75.

SECTION 4
Quantitative Reasoning

Time—26 minutes

15 Questions

Question 1

Effect of a New Fertilizer on Crop Yield

Effect of Fertilizer Application	Number of Plots
Significant increase in crop yield	28
No significant increase or decrease in crop yield	17
Significant decrease in crop yield	5
Total	50

Quantity A	Quantity B
The percent of plots that did not show a significant increase in crop yield	35%

A. Quantity A is greater.
B. Quantity B is greater.
C. The two quantities are equal.
D. The relationship cannot be determined from the information given.

Question 2

$N = 11121314\cdots50$

The integer N is formed by writing the consecutive integers from 11 through 50, from left to right.

Quantity A	Quantity B
The 26th digit of N, counting from left to right	The 45th digit of N, counting from left to right

A. Quantity A is greater.
B. Quantity B is greater.
C. The two quantities are equal.
D. The relationship cannot be determined from the information given.

Question 3

Six more than $\frac{1}{2}$ of the number r equals 14.

Three fewer than the square root of the number w equals 1.

Quantity A	Quantity B
r	w

A. Quantity A is greater.
B. Quantity B is greater.
C. The two quantities are equal.
D. The relationship cannot be determined from the information given.

Question 4

p is a prime number greater than 3.

Quantity A	Quantity B
The number of positive divisors of 2p	The number of positive divisors of p^2

A. Quantity A is greater.
B. Quantity B is greater.
C. The two quantities are equal.
D. The relationship cannot be determined from the information given.

Question 5

In the xy-plane, a line that has a slope of −3 passes through the points (3, k) and (−2, m).

Quantity A	Quantity B
k − m	−15

A. Quantity A is greater.

B. Quantity B is greater.

C. The two quantities are equal.

D. The relationship cannot be determined from the information given.

Question 6

Two water faucets are used to fill a certain tank. Running individually at their respective constant rates, these faucets fill the empty tank in 12 minutes and 20 minutes, respectively. If no water leaves the tank, how many minutes will it take for both faucets running simultaneously at their respective rates to fill the empty tank?

A. 16.0   B. 10.5   C. 8.0   D. 7.5   E. 6.0

Question 7

On the first day of work for a certain task, $\frac{1}{3}$ of the work was completed. On the second day of work for the task, $\frac{1}{4}$ of the work that remained was completed. What fraction of all the work for the task remained to be completed after the second day?

A. $\frac{1}{6}$ B. $\frac{1}{4}$ C. $\frac{5}{12}$ D. $\frac{1}{2}$ E. $\frac{7}{12}$

Question 8

A gardener plans to cover a rectangular plot of land with pine bark mulch to a depth of 4 inches. The plot measures 8 feet by 12 feet, and the gardener will buy mulch packed in bags. If each bag contains 3.5 cubic feet of mulch and costs $6, what is the cost of the least number of bags that the gardener will need to cover the plot? (Note: 1 foot = 12 inches.)

A. $30   B. $48   C. $54   D. $55   E. $60

Question 9

If x and y are positive integers and $x + y = 8x + 22$, which of the following must be true?

A. x is even.   B. xy is odd.   C. $x - y$ is odd.

D. $x(y+1)$ is even.   E. x and y are both odd.

Question 10

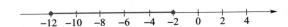

The solution set of which of the following inequalities is graphed on the number line shown?

A. $|x| \leqslant 12$ B. $2 \leqslant |x| \leqslant 12$ C. $|x-2| \leqslant 10$

D. $|x+7| \leqslant 5$ E. $|x-7| \leqslant 5$

Question 11

In the cube shown, the length of diagonal AB is 6. What is the surface area of the cube?

A. 24   B. 36   C. 54
D. 72   E. 108

Question 12

What is the remainder when $(345,606)^2$ is divided by 20?

_____

Question 13

What is the greatest number of nonoverlapping regions (that is, regions having no interior points in common) into which a rectangular region can be divided by three straight lines?

A. Five   B. Six   C. Seven   D. Eight   E. Nine

Question 14

If x is the average (arithmetic mean) of 10 and y, and y is the average of x and 22, which of the following statements are true?

Indicate <u>all</u> such statements.

A. $x < y$

B. $y - 10 = 22 - x$

C. The average of x and y is 16.

Question 15

In the four quarters of 2013, denoted by Q1, Q2, Q3, and Q4, company C hired the same number of employees in Q2 as in Q1 and twice as many employees in Q3 as in Q2. The number of employees hired by the company in Q4 was greater than the number hired in Q3; however, the number hired in Q4 was also less than 3 times the number hired in Q3. All of the employees were hired only once.

If an employee is to be selected at random from all the employees hired during the four quarters, which of the following values could be the probability that the employee will be one who was hired in Q4? Indicate all such statements.

A. $\dfrac{1}{3}$ B. $\dfrac{3}{8}$ C. $\dfrac{5}{12}$ D. $\dfrac{1}{2}$

E. $\dfrac{11}{20}$ F. $\dfrac{3}{5}$

模考二

SECTION 1
Quantitative Reasoning

Time—21 minutes

12 Questions

Question 1

$(n-2)(n-3) = 0$

Quantity A	Quantity B
$\left(\dfrac{n+1}{n}\right)^n$	2

A. Quantity A is greater.

B. Quantity B is greater.

C. The two quantities are equal.

D. The relationship cannot be determined from the information given.

Question 2

$72 = 2^x(3^y)$

x and y are positive integers.

Quantity A	Quantity B
x	y

A. Quantity A is greater.

B. Quantity B is greater.

C. The two quantities are equal.

D. The relationship cannot be determined from the information given.

Question 3

n is an integer and $(n-5)^{9-n} = 1$

Quantity A	Quantity B
n	4

A. Quantity A is greater.

B. Quantity B is greater.

C. The two quantities are equal.

D. The relationship cannot be determined from the information given.

Question 4

$a(2b+6) = 0$

Quantity A	Quantity B
a	b

A. Quantity A is greater.

B. Quantity B is greater.

C. The two quantities are equal.

D. The relationship cannot be determined from the information given.

Question 5

Advertising Expenditures

Corporation	Expenditure
A	$ 2, 957, 000
B	$ 2, 462, 000
C	$ 3, 997, 000
D	$ 2, 458, 000
E	$ 2, 197, 000
F	$ 2, 687, 000
G	$ 3, 920, 000

The table shows the advertising expenditures of seven major corporations last year. What is the median of the seven expenditures?

$ _____

Questions 6-8 are based on the following data.

Sources of College Information—Survey of 2,400 High School Students

Source of Information	Percent Citing Source
Campus visit	55%
College alumni	15%
College catalog	28%
College ranking	20%
College Website	25%
Current students	30%
Family members	43%
High school counselors	32%
High school teachers	17%

6. How many of the sources of information listed in the table were cited by more than 700 students?

   A. Two B. Three C. Four D. Five E. Six

7. Of the students surveyed who cited a college Website as a source of information, 65 percent visited one to five college Websites, 25 percent visited six to ten college Websites, and the rest visited more than ten college Websites. How many students surveyed who cited a college Website visited more than ten college Websites?

   _____ students

8. Of the students surveyed, if 12 percent cited both high school teachers and family members as a source of information, what percent cited neither of those sources?

   A. 40% B. 44% C. 48% D. 52% E. 56%

Question 9

If $y - 2x = 1$ and $3y + z = 2$, what is the value of x for which $3y + 2z = 0$?
Give your answer as a fraction.
_____.

Question 10

Square $BDFG$ is inscribed in isosceles triangle ACE. If the area of triangular region ACE is 1, what is the area of triangular region BCD?

A. $\frac{1}{4}$ B. $\frac{1}{5}$ C. $\frac{1}{6}$

D. $\frac{1}{8}$ E. $\frac{1}{9}$

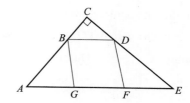

Question 11

An item with a suggested retail price of x dollars is being sold at three different stores. The selling prices at the three stores are 10 percent less, 15 percent less, and 20 percent less, respectively than the suggested retail price of the item.

Which of the following statements individually provides sufficient additional information to determine the value of x?

Indicate all such statements.

A. The range of the three selling prices is $10.
B. Only two of the three selling prices are greater than $80.
C. The average (arithmetic mean) of the three selling prices is $85.

Question 12

One student is to be selected at random from a class. The probability that the student selected will be male is equal to the probability that the student selected will be an English major. The probability that the student selected will be both male and an English major is 0.35. The probability that the student selected will be neither male nor an English major is 0.15. What is the probability that the student selected will be an English major?

A. 0.40   B. 0.45   C. 0.50   D. 0.55   E. 0.60

SECTION 2

Quantitative Reasoning

Time—26 minutes

15 Questions

Question 1

An electronics store purchased n television remote controls for x dollars each and sold each of them for $2.45

more than its purchase price. For the n remote controls, the total purchase price was $330 and the total of the selling prices was $477.

Quantity A	Quantity B
n	60

A. Quantity A is greater.
B. Quantity B is greater.
C. The two quantities are equal.
D. The relationship cannot be determined from the information given.

Question 2

x is a nonzero integer.

Quantity A	Quantity B
$(2^{-1})^x$	$\dfrac{1}{2x}$

A. Quantity A is greater.
B. Quantity B is greater.
C. The two quantities are equal.
D. The relationship cannot be determined from the information given.

Question 3

$|x| < 1 - x$

Quantity A	Quantity B
x	0

A. Quantity A is greater.
B. Quantity B is greater.
C. The two quantities are equal.
D. The relationship cannot be determined from the information given.

Question 4

List A consists of r integers, and list B consists of t integers, where $t = 3r$. The average (arithmetic mean) of the integers in the list A is 20, and the average of the integers in list B is 16.

Quantity A	Quantity B
The average of the $r + t$ integers in lists A and B	17.5

A. Quantity A is greater.
B. Quantity B is greater.
C. The two quantities are equal.
D. The relationship cannot be determined from the information given.

Question 5

Two shaded square regions, including their edges, are shown in the xy-plane and labeled I and II, respectively. S is the set of all possible slopes of line segments PQ where point P is in region I and point Q is in region II.

Quantity A
The greatest member of set S

Quantity B
$\frac{4}{3}$

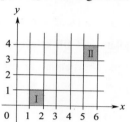

A. Quantity A is greater.
B. Quantity B is greater.
C. The two quantities are equal.
D. The relationship cannot be determined from the information given.

Question 6

If $n = ab$, where a and b are different prime numbers, which of the following statements must be true? Indicate all such statements.

A. $\sqrt{n}$ is an integer.
B. $n + 1$ is a prime number.
C. n^2 has 9 different positive factors.

Question 7

Printer P prints at the constant rate of x seconds per page, and printer Q prints at the constant rate of y seconds per page. If both printers print at their respective rates, then, in terms of x and y, what is the total number of pages the two printers print in 1 hour?

A. $\frac{3600}{x} + \frac{3600}{y}$

B. $\frac{x}{3600} + \frac{y}{3600}$

C. $\frac{3600}{x + y}$

D. $\frac{3600}{xy}$

E. $\frac{xy}{3600}$

Question 8

A young tree was planted on the first day of spring 2000. From the first day of spring 2000 to the first day of spring 2001, the height of the tree increased by 20 percent. From the first day of spring 2000 to the first day of spring 2002, the height of the tree increased by 38 percent. What was the percent increase in the height of the tree from the first day of spring 2001 to the first day of spring 2002?

A. 12% B. 15% C. 18% D. 24% E. 25%

Question 9

A bag contains 32 identically wrapped pieces of candy, of which 7 are mints, 11 are caramels, and the rest are chocolates. If Jill selects pieces of candy at random, what is the least number she must remove from the bag to be sure that she gets a caramel?

A. 11 B. 12 C. 19 D. 21 E. 22

Question 10

$N = 32^{19} - 32$

What is the units digit of N?

A. 1 B. 2 C. 4 D. 6 E. 8

Question 11

If n, k, and r are positive integers such that $n^k = 10r + 3$, which of the following could be the value of n?

A. 11 B. 12 C. 15 D. 17 E. 19

Question 12

In quadrant I of the xy-plane, a triangle has vertices at (r, r), $(4r, r)$, and $(r, 3r)$. For what value of r is the area of the triangle equal to 300?

_____

Question 13

In triangle ABC, point D lies on side AB so that AD is twice as long as DB. Point E lies on side BC so that DE

is parallel to *AC*. If the length of *DE* is 5, what is the length of *AC*?

A. 7   B. 9   C. 10   D. 12   E. 15

Question 14

In the figure, *ARD* is an arc of a circle centered at point *O*, *ABCD* is a rectangle, and $CD = \dfrac{a}{2}$. What is the area of the shaded region?

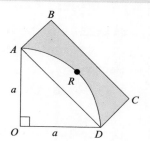

A. $\left(\dfrac{\sqrt{2}}{2} - \dfrac{1}{2}\right)a^2$

B. $\left(\dfrac{\sqrt{2}}{2} - \dfrac{\pi}{4}\right)a^2$

C. $\left(\dfrac{\sqrt{2}}{2} - \dfrac{\pi}{4} - \dfrac{1}{2}\right)a^2$

D. $\left(\dfrac{\sqrt{2}}{2} + \dfrac{\pi}{4} - \dfrac{1}{2}\right)a^2$

E. $\left(\dfrac{\sqrt{2}}{2} - \dfrac{\pi}{4} + \dfrac{1}{2}\right)a^2$

Question 15

All of the 80 science students at a certain school are enrolled in at least one of three science courses: biology, chemistry, and physics. There are 60 students enrolled in biology, 50 students enrolled in chemistry, and 35 students enrolled in physics. None of the students are enrolled in all three courses. Which of the following could be the number of students enrolled in both chemistry and physics?

Indicate all such numbers.

A. 0   B. 5   C. 10   D. 15   E. 20
F. 25   G. 30   H. 35

答案

▶ 模考一

Section 1
1-5 C/C/B/A/C   6-8 B/A/E/   9-12 12/E/E/B

Section 2
1-5 A/C/C/A/C   6-8 D/D/E   9-11 D/D/D   12-15 16/C/ABC/BCDE

▶ 模考二

Section 1
1-5 A/A/A/D/2687000   6-8 C/60/D   9-12 $\frac{1}{6}$/E/AC/E

Section 2
1-5 C/D/D/B/C   6-8 C/A/B   9-11 E/D/D   12-15 10/E/E/BCDE

附录

附录一　GRE 数学英文表达及词汇

1. 数学符号

equal to, the same as, is　等于（=）
more than　大于（>）
less than　小于（<）
no less than　大于等于（≥）
no more than　小于等于（≤）
add A to B, plus　加（+）
sum, total　结果（和）
minus, differ, subtract A from B　减（−）
difference　结果（差）
multiply　乘（×）
product　结果（乘积）
A is divided by B　除（÷）
dividend　被除数
quotient　商
rational number　有理数
irrational number　无理数
real number　实数
nonzero　非零
reciprocal　倒数
absolute value　绝对值（|⋯|）
square　平方（x^2）
cube　立方（x^3）
square root　开平方（$\sqrt{\square}$）
cube root　开立方（$\sqrt[3]{\square}$）

2. 数论

odd number　奇数
even number　偶数
positive number　正数
negative number　负数
integer　整数
factor/divisor　因数
multiple　倍数
prime number　质数
composite number　合数
prime factor　质因数
common factor/divisor　公因数
common multiple　公倍数
greatest common factor　最大公约数
least common multiple　最小公倍数
consecutive integers　连续整数
remainder　余数
decimal　小数
fraction　分数
decimal notation　十进制
decimal point　小数点
numerator　分子
denominator　分母
digit　数位上的数值
hundreds digit　百位数

tens digit　十位数
units/ones digit　个位数
tenths digit　十分位
2-digit number　两位数
round to, to the nearest　四舍五入
round up　只入不舍
round down　只舍不入
scientific decimal notation　科学计数法
terminating decimal　有限小数
repeating decimal　循环小数
non-repeating decimal　无限不循环小数
exponent　指数
base　底
power　幂
in terms of　用……表达
the number of　……的个数
exclusive　不包括端点值/开区间
inclusive　包括端点值/闭区间

3. 代数

equation　方程
variable　变量
constant　常数
coefficient　系数
solution　解
inequality　不等式
arithmetic sequence　等差数列
geometric sequence　等比数列
set　集合
subset　子集
sequence　序列
term　序列中的项
function　函数
linear function　线性函数

quadratic function　二次函数
domain　定义域
arithmetic mean　算术平均数
average　平均数
base　底数
closest approximation　近似
factorial　阶乘
maximum　最大值
minimum　最小值
per capita　人均

4. 几何

point　点
line　线
angle　角
degree　度数
acute angle　锐角
right angle　直角
obtuse angle　钝角
be parallel to　平行
be perpendicular to　垂直
polygon　多边形
triangle　三角形
quadrilateral　四边形
pentagon　五边形
hexagon　六边形
heptagon　七边形
octagon　八边形
nonagon　九边形
decagon　十边形
regular polygon　正多边形
equilateral triangle　等边三角形
isosceles triangle　等腰三角形
right triangle　直角三角形

hypotenuse 斜边
leg 侧边
rectangle 矩形
square 正方形
parallelogram 平行四边形
trapezoid 梯形
rhombus 菱形
perimeter 周长
area 面积
diagonal 对角线
altitude 高
width 宽
height 高
face 面
length 长度
dimension 大小，维度
distance 距离
angle bisector 角平分线
bisect 平分
circle 圆
center 圆心
radius 半径
diameter 直径
circumference 圆周长
chord 弦
arc 弧
sector 扇形
concentric circle 同心圆
circumscribe 外接
inscribe 内切
clockwise 顺时针
counterclockwise 逆时针
congruent 全等的
cube 正方体

rectangular solid 长方体
cylinder 圆柱
sphere 球
cone 圆锥
prism 棱柱
surface area 表面积
volume 体积
segment 线段
tangent 相切
opposite angle 对角
vertical angle 对顶角
congruent angle 全等角
vertices/vertex 顶点
intersect 相交
midpoint 中点
number lines 数轴
plane 平面
rectangular coordinate system 平面直角坐标系
quadrant 象限
origin 原点
coordinate 坐标
slope 斜率
intercept 截距
reflection of p about x-axis p 点关于 x 轴的对称点

5. 数据分析

principal/initial amount 本金
compound interest 复利
simple interest 单利
cost 成本
discount 折扣
interest rate 利率
interest 利息
list price 标价

margin   利润

mark up   涨价

mark down   降价

markup   毛利

profit   利润

revenue   收益

purchasing price   购买价

retail value   零售价

sale price   销售价

combination   组合

permutation   排列

probability/possibility   概率

independent events   独立事件

exclusive events   互斥事件

mean/average   平均数

median   中位数

mode   众数

range   极差

variance   方差

standard deviation   标准差

percentile   百分位数

quartile   四分位数

interquartile range   四分位距

normal distribution   正态分布

statistics   统计

frequency/count   频率

relative frequency   相对频率

box plot   箱线图

weighted mean   加权平均数

intersection   交集

union   并集

disjoint/mutually exclusive   互斥

greatest possible value   最大可能值

least possible value   最小可能值

附录二　GRE数学常用公式

1. 三角形不等式：$|a+b| \leq |a| + |b|$
2. 点(x_1, y_1)与点(x_2, y_2)的两点间距离公式$=\sqrt{(x_2-x_1)^2+(y_2-y_1)^2}$
3. 一元二次方程$ax^2+bx+c=0$的求根公式：
$$x=\frac{-b\pm\sqrt{b^2-4ac}}{2a}$$
4. 二次函数的对称轴：$x=-\frac{b}{2a}$
5. 平方差公式：$a^2-b^2=(a-b)(a+b)$
6. 完全平方公式：$(a+b)^2=a^2+2ab+b^2$，$(a-b)^2=a^2-2ab+b^2$
7. 等差数列的通项公式：$a_n=a_1+(n-1)d$
8. 等差数列的求和公式：$S_n=\frac{(a_1+a_n)n}{2}$
9. 等比数列的通项公式：$a_n=a_1q^{n-1}$
10. 等差数列的求和公式：$S_n=\frac{a_1(1-q^n)}{1-q}$
11. n边形内角和$=(n-2)\times 180°$
12. 底边a、高为h的三角形面积$=\frac{1}{2}ah$
13. 长为a、宽为b的矩形面积$=ab$
14. 底边为a、高为h的平行四边形面积$=ah$
15. 菱形的面积$=\frac{\text{两条对角线的乘积}}{2}$
16. 梯形的面积$=\frac{(\text{上底}+\text{下底})}{2}\times\text{高}$
17. 半径为r、直径为d的圆的周长$=2\pi r=\pi d$
18. 半径为r、直径为d的圆的面积$=\pi r^2=\frac{d^2}{4}\pi$
19. 弧长公式$=\frac{\text{圆心角}}{360°}\pi d=\frac{\text{圆心角}}{180°}\pi r$
20. 扇形面积$=\frac{\text{圆心角}}{360°}\pi r^2$
21. 边长为a的正方体的表面积$=6a^2$
22. 边长为a的正方体的体积$=a^3$
23. 长为l、宽为w、高为h的长方体的表面积$=2(wl+lh+wh)$
24. 长为l，宽为w，高为h的长方体的体积$=l\times w\times h$
25. 半径为r，高为h的圆柱体的表面积$=2\pi r^2+2\pi rh$
26. 半径为r、高为h的圆柱体的体积$=\pi r^2 h$
27. $n!=n\times(n-1)\times\cdots\times 2\times 1$
28. 组合$C_m^n=\frac{m!}{n!(m-n)!}$
29. 排列$P_m^n=\frac{m!}{(m-n)!}$